Embryology of Plants

Embryology of Plants

Rajesh Singh

Delve
PUBLISHING
www.delvepublishing.com

Embryology of Plants

Rajesh Singh

Delve Publishing
224 Shoreacres Road
Burlington, ON L7L 2H2
Canada
www.delvepublishing.com

Email: orders@arclereducation.com

ISBN: 978-1-77361-059-7 (Hardcover)

ABOUT THE AUTHOR

Prof Rajesh Singh, currently working as Vice Chancellor of Purnea University, Purnea, Bihar & Vice Chancellor of BBA Bihar University Muzaffarpur, Bihar . Earlier Prof Singh joined as Professor of Genetics and Plant Breeding since November 2006. He has over 29 years experience in teaching, research and extension in Central University (BHU), State University (GBPUAT, Pantnagar) as well as Central Institutions (Indian Council of Agriculture Research, New Delhi). He has been holding important charges of Coordinator/Head of Plant Biotechnology, Chairman Press Publication and Publicity Cell of BHU (equal to dean), Secretary-cum-Coordinator of a 50 million dollars International Project "Agriculture Innovation Partnership" funded by US AID (American Embassy) involving five American and three Indian universities, BHU being leading University. Dr Singh has taught for over 25 years at under-graduate and postgraduate levels in BHU, Varanasi and GBPUT, Pant Nagar and guided over 60 students of Master's and Doctoral level research so far. He has successfully generated external funding to the tune of about Rs 70 crores through 25 competitive grant external projects of national and international importance. He has developed and released 16 crop varieties/patent. He planned and executed 12 explorations for collection and conservation of about 4000 multi-crop germplasm. He has accepted consultancy of establishing DNA finger printing labs of UP and Bihar state government and was also pioneer in successfully establishing five new biotechnology labs. Dr Singh was instrumental in reforming / initiating several courses in Genetics, Molecular Biology, Molecular

Breeding, Plant Breeding, Cyto-genetics, Microbial Genetics, etc including, successfully starting a new course of M.Sc (Plant Biotechnology) by planning, preparing and executing the course curriculum as per guide lines of UGC and ICAR. Prof Singh has been conferred with several awards including SEE Fellow Award, Best Scientist cum Field Worker Award, Kisan Sewa Award, Young Plant Biotechnologist Award, Best Scientist Award, Innovation Award, Excellence Award, etc. He has published about 250 publications including 105 peer reviewed research papers and 21 books/bulletin and manuals of national and international importance.

TABLE OF CONTENTS

GLOSSARY

A

Aberrant – departing from an accepted standard

Abiotic – physical rather than biological; not derived from living organisms

Accumulates – gather together or acquire an increasing number or quantity of

Accumulation - The acquisition or gradual gathering of something

Acrosome - is an organelle that develops over the anterior half of the head in the spermatozoa

Adaptation – The process of change by which an organism or species becomes better suited to its environment

After-ripening - The period of internal change that is necessary in some apparently mature seeds before germination can occur

Alleles – each of two or more alternative forms of a gene that arise by mutation and are found at the same place on a chromosome

Amyloplasts - Amyloplasts are a type of plastid, double-enveloped organelles in plant cells that are involved in various biological pathways.

Analysis – Detailed examination of the elements or structure of something

Angiosperm - a plant of a large group that comprises those that have flowers and produce seeds enclosed within a carpel, including herbaceous plants, shrubs, grasses, and most trees

Apical end - The fraction of the cell membrane at the apical end of the cell, which faces the outside world or the lumen of the cavity

Apomixis - In botany, apomixis was defined by Hans Winkler as replacement of the normal sexual reproduction by asexual reproduction, without fertilization.

Appendage - A projecting part of an invertebrate or other living organism, with a distinct appearance or function.

Arabidopsis thaliana - the thale cress, mouse-ear cress, or Arabidopsis, is a small flowering plant native to Eurasia and Africa

Arche – beginning

Asymmetrically - having parts which fail to correspond to one another in shape, size, or arrangement; lacking symmetry

Attributed – regard something as being caused by

Autonomously - with the freedom to act independently

Autotrophic - an organism that can produce its own food using light, water, carbon dioxide, or other chemicals

Auxotrophic – Designating or of a mutant organism requiring more nutritional substances than its prototrophic parent because it has lost the ability to make a certain enzyme

B

Basal Cell – found at the bottom of the epidermis — the outermost layer of skin

Bicellular – having or composed of two cells

Biosynthesis - The production of complex molecules within living organisms or cells

Biotic – relating to or resulting from living organisms

Bivalents - A pair of homologous chromosomes.

Bract – a modified leaf or scale, typically small, with a flower or flower cluster in its axil

Brassicas - a plant of a genus that includes cabbage, swede, rape, and mustard

Breeding - The mating and production of offspring

Bumping - knock or run into someone or something with a jolt

C

Chalazal - a structure inside bird and reptile eggs and plant ovules

Chimerism - A genetic chimerism or chimera is a single organism composed of cells with more than one distinct genotype.

Chloroplast – A plastid in green plant cells which contains chlorophyll and in which photosynthesis takes place

Cleavage - cell division, especially of a fertilized egg cell

Clonal - relating to or characteristic of a clone

Coexist – Exist at the same time or at the same place

Conjugation - the temporary union of two bacteria or unicellular organisms for the exchange of genetic material

Conquer – overcome and take control of (a place or people) by military force

Consistently –In every case or on every occasion; invariably

Consumption – The action of using up a resource

Contention – heated disagreement

Contestable – capable of being contested

Copious – abundant in supply or quantity

Corolla - the petals of a flower, typically forming a whorl within the sepals and enclosing the reproductive organs

Cotyledons – an embryonic leaf in seed-bearing plants, one or more of which are the first leaves to appear from a germinating seed

Counterparts - a person or thing that corresponds to or has the same function as another person or thing in a different place or situation

Crucifers - a vegetable of the mustard family: especially mustard greens; various cabbages; broccoli; cauliflower; brussels sprouts

Cultivars – a plant variety that has been produced in cultivation by selective breeding

D

Daunted - make (someone) feel intimidated or apprehensive

Degrading - Break down or deteriorate chemically

Dehiscent - characterized by splitting or bursting open

Deleterious – causing harm or damage

Desiccation - The removal of moisture from something

Determinacy - The quality or state of being determinate; the state of being definitely and unequivocally characterized: exactness.

Dicotyledonous - also known as dicots, are one of the two groups into which all the flowering plants or angiosperms were formerly divided

Dioecious - (of a plant or invertebrate animal) having the male and female reproductive organs in separate individuals.

Diploid - of a cell or nucleus) containing two complete sets of chromosomes, one from each parent

Dispersal - the action or process of distributing or spreading things or people over a wide area

Disrupted - Interrupt (an event, activity, or process) by causing a disturbance or problem

Disturbed – having had the normal pattern or functioning disrupted

Divergence - A situation in which two things become different, or the difference between them increases.

Dormancy - a vegetable of the mustard family: especially mustard greens; various cabbages; broccoli; cauliflower; brussels sprouts

Dormancy – The state in which a plant is alive but not actively growing

Drab – lacking brightness or interest

Drier – free from moisture or liquid; not wet or moist

E

Efflux - they are active transporters, meaning that they require a source of chemical energy to perform their function

Elucidating – make (something) clear; explain

Emasculation - is the removal of both the penis and the testicles, the external male sex organs

Embedded – (of an object) fixed firmly and deeply in a surrounding mass; implanted

Embryo – A thing at a rudimentary stage that shows potential for development

Embryo Proper – the entire embryo exclusive of the suspensor

Embryology - the branch of biology that studies the prenatal development of gametes, fertilization, and development of embryos and fetuses

Encoding - (of a gene) be responsible for producing (a substance or behaviour)

Endosperm - The part of a seed which acts as a food store for the developing plant embryo, usually containing starch with protein and other nutrients

Enormously – to a very great degree or extent

Exocytosis – a process by which the contents of a cell vacuole are released to the exterior through fusion of the vacuole membrane with the cell membrane

Exogenous - Growing or originating from outside an organism

F

Fertilization - the action or process of fertilizing an egg or a female animal or plant, involving the fusion of male and female gametes to form a zygote

Flavonoids - Any of a large class of plant pigments having a structure based on or similar to that of flavone

Flounder – struggle or stagger clumsily in mud or water

G

Gamete - a mature haploid male or female germ cell which is able to unite with another of the opposite sex in sexual reproduction to form a zygote

Geitonogamy - The fertilization of a flower by pollen from another flower on the same (or a genetically identical) plant.

Gemmae - a single cell, or a mass of cells, or a modified bud of tissue, that detaches from the parent and develops into a new individual

Genotypes - The genetic constitution of an individual organism

Germinating – begin to grow and put out shoots after a period of dormancy

Germination – The development of a plant from a seed or spore after a period of dormancy

Gibberellins - Gibberellins (GAs) are plant hormones that regulate various developmental processes, including stem elongation, germination, dormancy, flowering, flower development, and leaf and fruit senescence.

Globular – globe-shaped; spherical

Gnetales - some botanists to be the ancestors of flowering plants or angiosperms

Gymnosperms - also known as Acrogymnospermae, are a group of seed-producing plants that includes conifers, cycads, Ginkgo, and gnetophytes

H

Haploid - (of a cell or nucleus) having a single set of unpaired chromosomes

Hermaphroditism - The condition of having both male and female reproductive organs.

Herring – are forage fish, mostly belonging to the family Clupeidae

Heterozygosity - The condition of having two different alleles at a locus; It is fundamental to the study of genetic variation in populations.

Heterozygous – having two different alleles of a particular gene or genes

Hindering – make it difficult for (someone) to do something or for (something) to happen

Homozygosity - The possession of two identical alleles of a particular gene or genes by an individual.

Hormones – A regulatory substance produced in an organism and transported in tissue fluids such as blood or sap to stimulate specific cells or tissues into action

Hypophysis - a small, pea-sized gland located at the base of the brain that functions as «The Master Gland.»

I

Idiosyncratic - An unusual feature of a person; relating to odd habit.

Impermeable - Not allowing fluid to pass through

Inducing – Bring about or give rise to

Inflorescence - An inflorescence is a group or cluster of flowers arranged on a stem that is composed of a main branch or a complicated arrangement of branches.

Influx – an arrival or entry of large numbers of people or things

Inherent – Existing in something as a permanent, essential or characteristic attribute

Inhibition – The slowing or prevention of a process, reaction or function by a particular substance

Integuments - a tough outer protective layer, especially that of an animal or plant

Intersecting - (of two or more things) pass or lie across each other

Intraspecific - produced, occurring, or existing within a species or between individuals of a single species

L

Lethality – the capacity to cause death or serious harm or damage

Life-cycle – The series of changes in the life of an organism including reproduction

M

Markedly – to an extent which is clearly noticeable; significantly

Maturation – The action or process of maturing

Meiosis - a type of cell division that results in four daughter cells each with half the number of chromosomes of the parent cell, as in the production of gametes and plant spores

Membranous – thin, pliable, and often somewhat transparent

Meristem - a region of plant tissue, found chiefly at the growing tips of roots and shoots and in the cambium, consisting of actively dividing cells forming new tissue

Metabolic - Relating to or deriving from the metabolism of a living organism

Meticulous - Showing great attention to detail; very careful and precise.

Microarray - A set of DNA sequences representing the entire set of genes of an organism, arranged in a grid pattern for use in genetic testing

Micropylar - A minute opening in the ovule of a seed plant through which the pollen tube usually enters

Micropyle – a small opening in the surface of an ovule, through which the pollen tube penetrates, often visible as a small pore in the ripe seed

Microtubules - Microtubules are polymers of tubulin that form part of the cytoskeleton and provide structure and shape to eukaryotic cells.

Molecular biology - Molecular biology is the branch of biology that concerns the molecular basis of biological activity in and between cells, including molecular synthesis, modification, mechanisms and interactions.

Monocotyledonous - commonly referred to as monocots, are flowering plants, the seeds of which typically contain only one embryonic leaf, or cotyledon

Morphogenesis - is the biological process that causes an organism to develop its shape

Morphogenic - relating to the origin and development of morphological characteristics

Mosses – a small flowerless green plant which lacks true roots, growing in low carpets or rounded cushions in damp habitats and reproducing by means of spores released from stalked capsules

Musky – having a smell or taste like that of musk

Mutant – Resulting from or showing the effect of mutation

N

Nematodes - The nematodes or roundworms constitute the phylum Nematoda, with plant-parasitic nematodes being known as eelworms.

Niche - a shallow recess, especially one in a wall to display a statue or other ornament

Nocturnally – relating to the night

Non-Motile - Not exhibiting or capable of movement.

Nucleus - The central and most important part of an object, movement, or group, forming the basis for its activity and growth; a dense organelle present in most eukaryotic cells, typically a single rounded structure bounded by a double membrane, containing the genetic material.

O

Obscured – keep from being seen; conceal

Outcrossed – breed (an animal or plant) with one not closely related

Overexpressed - To make too many copies of a protein or other substance

P

Paleontological - Paleontology is the study of the history of life on Earth as based on fossils.

Parietal - Relating to or denoting the wall of the body or of a body cavity or hollow structure.

Pathway – A way of achieving a specified result; a course of action

Penetration - the action or process of penetrating something

Periclinal - (of a cell wall) parallel to the surface of the meristem

Perivitelline – is the space between the zona pellucida and the cell membrane of an oocyte or fertilized ovum

Phenotypes – The set of observable characteristics of an individual resulting from the interaction of its genotype with the environment

Phylogenetic - relating to the evolutionary development and diversification of a species or group of organisms, or of a particular feature of an organism

Physiological – Relating to the branch of biology that deals with the normal functions of living organisms and their parts

Pigmentation - The natural coloring of animal or plant tissue

Pistil - the female organs of a flower, comprising the stigma, style, and ovary

Placenta - a flattened circular organ in the uterus of pregnant eutherian mammals, nourishing and maintaining the fetus through the umbilical cord

Plasmodesmata - are microscopic channels which traverse the cell walls of plant cells and some algal cells, enabling transport and communication between them

Pollen - Pollen is a powdery substance consisting of pollen grains which are male microgametophytes of seed plants, which produce male gametes (sperm cells).

Polyphilic - The term polyphilic is less common and was originally coined in the field of liquid crystals.

Precocious - Having developed certain abilities or inclinations at an earlier age than is usual or expected.

Pre-requisite – a thing that is required as a prior condition for something else to happen or exist

Primordium - An organ, structure, or tissue in the earliest stage of development.

Proboscis - The nose of a mammal, especially when it is long and mobile such as the trunk of an elephant or the snout of a tapir.

Procambium - the part of the primary meristem of a plant that forms cambium and primary vascular tissues

Propagules - any material that functions in propagating an organism to the next stage in its life cycle, such as by dispersal

Protoderm - The primary meristem in vascular plants that gives rise to epidermis. Also called dermatogen

Protogyny - The condition of flowers whose female parts mature before the male ones.

Proximal - Situated nearer to the centre of the body or the point of attachment.

Proximity – nearness in space, time, or relationship

Purged - rid (someone or something) of an unwanted quality, condition, or feeling

Q

Quiescence - inactivity or dormancy

R

Radicle - The part of a plant embryo that develops into the primary root

Recalcitrant – having an obstinately uncooperative attitude towards authority or discipline

Recessive – relating to or denoting heritable characteristics controlled by genes which are expressed in offspring only when inherited from both parents

Recombination – the process of recombining things

Reminiscent – tending to remind one of something

Reserves – A supply of a commodity not needed for immediate use but available if required

Rhizoid - also referred to as microfilaments and are highly conserved, proteinous components found near ubiquitously in eukaryotic cytoskeletons

Ridge – a long, narrow hilltop, mountain range, or watershed

Rudimentary - involving or limited to basic principles

Runner Beans - Phaseolus coccineas, known as runner bean, scarlet runner bean, or multiflora bean, is a plant in the legume family, Fabaceae

Rupture - Cause to break or burst suddenly

S

Scars - a mark left on the skin or within body tissue where a wound, burn, or sore has not healed completely and fibrous connective tissue has developed

Seedling – A young plant, especially one raised from seed and not from a cutting

Segregation – the action or state of setting someone or something apart from others

Severed - having been cut or sliced off

Spatiotemporal – belonging to both space and time or to space–time

Spindles - a slender rounded rod with tapered ends used in hand spinning to twist and wind thread from a mass of wool or flax held on a distaff

Spore - A minute, typically one-celled, reproductive unit capable of giving rise to a new individual without sexual fusion, characteristic of lower plants, fungi, and protozoans.

Sporophyte - in the life cycle of plants with alternating generations) the asexual and usually diploid phase, producing spores from which the gametophyte arises

Sterically – caused by the spatial arrangement of atoms in a molecule

Stimulate – Encourage development of or increased activity in (a state or process)

Stratification - The placing of seeds close together in layers in moist sand or peat to preserve them or to help them germinate

Suspensor cell - a suspending part or structure: such as. a : a group or chain of **cells** that is produced from the zygote of a seed plant and serves to push the developing embryo into the endosperm

Sustaining - cause to continue for an extended period or without interruption

Symbiotic - Symbiosis is any type of a close and long-term biological interaction between two different biological organisms, be it mutualistic, commensalistic, or parasitic.

Symplast - the inner side of the plasma membrane in which water and low-molecular-weight solutes can freely diffuse

Synapsis - The fusion of chromosome pairs at the start of meiosis.

Syncytial - A multinucleate mass of cytoplasm that is not separated into cells.

T

Taxa - a taxonomic group of any rank, such as a species, family, or class

Testa - The protective outer covering of a seed; the seed coat

Tetrad - a group or set of four

Tolerance - The capacity to endure continued subjection to something such as a drug or environmental conditions without adverse reaction

U

Ultrastructural – is the architecture of cells and biomaterials that is visible at higher magnifications than found on a standard optical light microscope

Undeciphered - (of speech or writing) not converted into normal language or understood

Unfavorable – Likely to lead to an adverse outcome

Unfolded - (of events or information) gradually develop or be revealed

V

Vacuole - A space or vesicle within the cytoplasm of a cell, enclosed by a membrane and typically containing fluid.

Variations - The occurrence of an organism in more than one distinct colour or form

Vesicle - A vesicle is a structure within or outside a cell, consisting of liquid or cytoplasm enclosed by a lipid bilayer.

Vigor - physical strength and good health

Voraciousness - To eat greedily; to devour.

W

Whorls - a pattern of spirals or concentric circles

Z

Zygote - a diploid cell resulting from the fusion of two haploid gametes; a fertilized ovum

LIST OF FIGURES

PREFACE

Embryology is the study of the prenatal development of the gametes. The study can be very helpful in the development of cells in the manner that scientists or researchers may want. Embryology of plants gives a deep insight into the way in which the plant cells develop and thus provide an extensive area of research for scientists and researchers.

This book aims to provide the readers with all aspects of the embryology of plants so that they can equip themselves with all the available and accessible information and can develop a good foundation on which they can conduct their studies and researches.

The book starts off by introducing the subject of plant embryology to the readers and explaining to them the way in which the embryo develops, including the major processes that take place. The second chapter then throws light on the reproduction process in plants and the related terminologies and aspects.

The third chapter explains the readers the process of fertilization in plants, while the fourth chapter talks about Gametogenesis, which basically involves the development of gametes. Then, the fifth chapter takes the readers through the processes of pollination and breeding in plants, explaining to them the various ways in which the pollination takes place.

The sixth chapter talks about endosperm and embryology and informs the readers about their aspects in detail. The seventh chapter throws light on the concept of seed dormancy and the process of seed germination and also informs about the significance these processes hold in plant embryology. The last chapter talks about the zygotic embryogenesis and takes the readers through all the aspects related to the topic.

With this book, I hope to inform the readers about plant embryology in detail. My effort is to make maximum knowledge available to the readers so that they can employ all such information to the best of their ability, in areas of research and development and even in basic educational purposes that pertain to their own curiosity about the subject.

Chapter 1 Introduction to Plant Embryology

CONTENTS

Plant Embryogenesis, a process that aims to develop a full plant from an embryo. It occurs after the fertilization of an ovule. This stage is followed by dormancy and germination in the plant life cycle. This chapter gives insights into the embryogenesis in plants. The first half of the chapter discusses the development of an embryo in dicots and monocots. Embryogenesis in higher plants is explained in the mid of the chapter followed by the morphological description of embryogenesis. The chapter ends with the discussion regarding the major process involved in embryonic development.

1.1. INTRODUCTION

The egg after the process of getting fertilized is called as zygote or oospore. It later develops into an embryo. Before the process actually starts the zygote enters into a stage where it undergoes A period of rest. This period where are you from few hours to a few months. Generally, there is an immediate division of the zygote after the first division of the primary endosperm nucleus.

There is always wall formation which results in a two-celled pro embryo. This happens after the first division of the zygote. The case is different in gymnosperms where there is free nuclear divisions at the early stages of their development.

However, in monocots and dicots, there is no fundamental difference in their early stages of development of the embryos. But it is noticed that in the later stages, the embryos of dicots and monocots show a marked difference. The embryo of dicotyledonous and monocotyledonous plants has been considered separately here in the process of embryogenesis.

1.2. DEVELOPMENT OF EMBRYO IN DICOTS

As per Soueges, the embryonal type is classified according to the contribution made by each of these cells that makes a base and the mode of origin of the four-celled pro-embryo. However, in dicotyledons, there are five main types of embryos that have been recognized by Schnarf (1929), Johansen (1945) and Maheshwari (1950).

1.2.1. The terminal cell of the two-celled pro-embryo divides by a longitudinal wall.

- Crucifer Type

There is very less or no role of the basal cell in the development of the embryo

- Asterad Type

The role of basal and terminal cells is important in the development of the embryo.

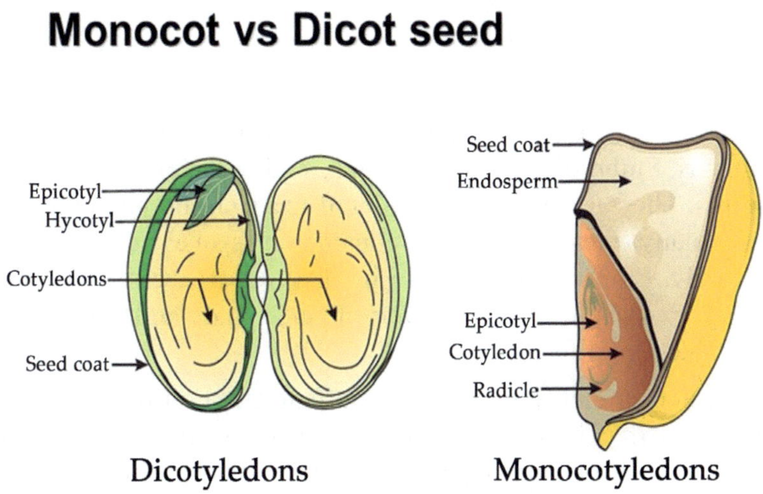

Figure 1.1: Monocot and dicot seed.

Source: Image by Quora

1.2.2. The terminal cell of the two-celled proembryo divides by a transverse wall, Basal cell plays a little or no role in the development of the embryo.

- Solanad Type

There is two or more suspensor cell formation in the basal cell.

- Caryophyllod Type

Here is a further division of the basal cell.

- Chenopodiad Type

The basal and terminal cells both take part in the development of the embryo.

1.3. DEVELOPMENT OF EMBRYO IN MONOCOTS

The only difference between the monocotyledons and the dicotyledons is the matured embryos. However, there is no significant difference between them regarding the early cell divisions of the proembryo.

> Embryonic development also termed as embryogenesis is a widespread phenomenon in plants which also consists of a part of our diet, for example, Grains, Beans, Nuts etc.

Embryology is that branch of biology which studies the development of new organisms. Embryology in plants, thus, studies the origin and formation of new plants. They are not only considered as a rich source of nutrition but also attracts many researchers as it is considered as an extremely interesting object of study. Embryos are diverse in structure as well as function and hence embryogenesis in plants is a very diverse study and is not limited to the only production of fertilization.

The first and probably the most important question here is that "what is a plant embryo?" The most common answer for this is that the embryo is an unborn new organism or unhatched new organism in its early stages of development. When it is understood in terms of sexually reproducing organisms, then it is the fertilization that adds to its property.

It is difficult to define here that what should be the correct definition of a plant embryo. A zygote is formed when the green algae conjugate with its sister land plant, charophycean lineage. It then gives rise to spores after undergoing meiosis. This is the process of embryogenesis, but here in the case of land plants, it should be known as embryophytes.

There has been an evolution of several plant line images for the purpose of vegetative propagation such as primer gemmae in liverworts.

The structure formed sometimes resembles zygotic embryos. Therefore, in plants, it is difficult to decide what actually an embryo is whether it is ontogeny or morphology. There is no good basis for deciding it. There are many ways that make an embryo.

Figure 1.2: Liverworts sometimes resemble a zygote.

Source; Image by Wikimedia Commons

1.3.1. Many Ways to Become an Embryo

There are many ways to make an embryo. The most important and defining step in this way is the initiation of the plant embryogenesis from a non-embryonic precursor cell. The haploid gamete is transformed into a diploid sporophyte. This event occurs naturally as the egg cell gets fertilized. This event leads to the reprogramming of the cellular state. This further the process of embryo development.

Figure 1.3: An example of ginkgo embryo.

Source: Image by Wikimedia Commons

Contrastingly, in plants, the process of the embryo initiation is not at all limited to the egg cell. In the late 1950s, it was hypothesized that there was the production of large numbers of embryos when carrot somatic cells were grown in a fully synthetic medium with the synthetic auxin phytohormone 2,4-D.

However, in many species of plants, this conditional reprogramming of somatic cells in the process of embryogenesis was found to be common in the cells of different origins. It was found to be able to be triggered by stresses in haploid microspores. Therefore, the process of embryogenesis can be universally triggered in the cells of plants. This raises the question of the way the process needs to be regulated.

The recent work in the way shows that the process and the reprogramming require chromatin level changes. It is found that the following process or step needs to remove the epigenetic marks that usually repress the embryogenic potential. Therefore, it is needed to be removed before the potential embryo is unleashed.

Similarly, fertilization induces dramatic changes in the chromatin structure and nuclear architecture. Thus, the process of conversion from the non-embryonic state to the embryonic state requires epigenetic reprogramming.

This reprogramming is not related to the origin of the embryo. There is the identification of several positive regulators of embryo initiation such as the transcription factors and the repressive chromatin control.

In a specific context, it can be said that there are several transcription factors that can induce the process of embryogenesis when it is overexpressed. However, there is no special evidence that links the initiation of the embryogenesis process with the function of the proteins in the normal context of its functions.

In a conceptual sense, there can be a competent and differentiated cell state if the repressive chromatin markers are removed. It can also be induced by the overexpression of some transcription factors in parts.

The competent cells can enter any new developmental program and also enter the embryogenesis process. It is there due to the changes in the hormones, stresses, and the aspects of cellular state.

At present, there is a lack of an experimental basis for the casual relationship among the competence, chromatin control and embryogenesis. The sequence must have been clear at the time. The plant cell has the ability to switch back to the embryogenesis. The question here arises that what led to the switching back to the embryogenesis process.

There cannot be reprogramming done for all the cells even for *in vitro* embryogenesis. The real fact of the matter is that only a small number of cells undergoes a complete programming in all the documented cases of embryogenic cell formation. There are different explanations for this heterogeneity.

Depending on the prior developmental state of cells, the competence for reprogramming could have underlying differences.

The epigenetic marks needed to be removed to some extent. The factor the decides the extent to which these marks are needed to be removed is the somatic cells having the variability in the cellular context in the tissue (epidermal cells or cambial stem-like cells).

However, it is totally dependent on the success of this method to remove the marks. Also, it is unknown whether it will remove the epigenetic marks selectively or partially. At later stages, the transition may fail fully.

The cell's ability to be reprogrammed depends on its interaction with other cells that surround it. This can be taken as the second explanation for the above discussed. The interactions that are observed are the inhibitory and stimulatory in the embryogenic cell cultures. Additionally, the interactions among the signal molecules have also been identified.

These include cytokinin, polyamines, peptides, complex proteoglycans, oligosaccharides, and enzymes. The interaction among the cells is suggested by certain spatial periodicity in responding cells in the basal plants.

For example, patterns reprogramming in lateral inhibition. There are chances in real that the current passes in the cells during reprogramming given that there is a continuous symplast through plasmodesmata connections in many of the plant tissues.

The lateral inhibition case starts in the neighboring cells. Like it has been proposed that there are numerous embryos of plant species of flowers that suppresses the process of embryogenesis in the neighboring suspensor cells.

It can be clearly said that the single- cell level needs some approaches that are required to predict and describe what the programming towards embryogenesis is. This would help to understand what is a definition of a plant embryo at a molecular level.

1.3.2. The Early Embryo as a Model for Plant Development

In order to initiate the process of making an embryo, the young plant has to face several tasks. Until the time the seed desiccates, which is done in a limited time period, there is a need for sufficient cells that will build a new body.

Moreover, the cell divisions are needed to be put in an ordered way that would help in the achievement of a morphology that is species-specific. Thereafter, as there is progress in the embryo, the stem cells, tissues, and various cell types and are needed to be defined.

Therefore, it can be seen that the process of embryogenesis is a period of intense growth, pattern formation and morphogenesis. There has been much learning from the Arabidopsis thaliana, which is basically a flowering plant.

The studies of this flowering plant species are having strong parallel to other flowering plant species. Here, the summarization is done for making the process of embryogenesis, clearer with the example of Arabidopsis plant.

Figure 1.4: *Arabidopsis thaliana*, a flowering plant.

Source: Image by Wikimedia Commons

One of the interesting properties of *Arabidopsis* is the cell division patterns in the embryo which are close to invariant.

Generally, this property is shared by the crucifer family. However, it is not representative of the flowering plants. The flowering plant *Arabidopsis* can be regarded as a minimalist. It generates an embryo with very few cells possible.

Therefore, each cell has a strong constraint with regard to the division patterns. However, this behavior may be idiosyncratic. The facile detection of the cell is allowed by this property and also provides help to the cell division for the developmental control through a good model.

1.3.3. Oriented and Asymmetric Cell Division

The ability of the cells to be differentiated from each other cell is the deciding factor in the development of the multicellular organisms. It should be clearly noted that even for producing a miniature plant from a single cell, the process of embryogenesis must produce all the cell types that a mature seedling embryo must have.

Therefore, it is clear from this discussion that an early embryo is the one that has the mechanism which generates the active cellular diversity.

Taking the above into consideration, an example of such a mechanism is n asymmetric cell division. The birth of two daughter cells having different properties marks the asymmetry of division. This could even generate two cells with different identities if it is in the context of the developmental state. The difference in the cells would be in terms of size and shape.

The embryo of Arabidopsis has 3D cell shapes, the systematic analysis of which says that there are different modes of asymmetry which are correlated too.

In the formation of the embryo in the case of Arabidopsis, there is a formation of two different cell types which have different volumes also. This is considered as the embryonic cell population known as the pro-embryo and the extra-embryonic lineage known as the suspensor.

In many cases, the proembryo is formed by the cell which has no connection to the seed tissue, the cell is the upper cell that develops into an embryo. So far, other than correlation, there has been no observation in this connection.

The question that is the most important to think is whether the first division asymmetry is casual to the distinctions between the embryonic and extra-embryonic cell lineage.

In this regard, the studies that are related to genetics in Arabidopsis are not unambiguous. Some mutants are present there to disturb the first division.

However, it should be noted that this does not eliminate the distinction between the two cell fates. The cellular execution and genetic regulation has been discussed further with respect to the first division.

There is a requirement of paternal input in the symmetry of the first division in Arabidopsis zygotic embryos. The zygote and asymmetric division get elongated by the membrane protein that is derived from the male gamete.

The consolidation of the intrinsic polarity of the egg cell is done before the fertilization starts. This consolidation is in view of this polarity and the asymmetric division of the polarity helps in the consolidation process.

However, this is not the case every time. As and when the fertilization occurs there is a loss of anatomical polarity in the zygote and is further repolarized. There is a requirement of a specific set of genes that requires a number of transcription factors in repolarization.

It is evident from the history that there are few and instruments present in plants that can control the process of morphogenesis in 3D and oriented cell division being one of them.

There is a set of patterns followed by the plant cell division which is approximated by simple rules in 2D. It was proposed in the mid -late 1800s. These rules are applied in some of the following examples such as the constant curvature of the new wall, the center of the cell having the shortest wall and the attachment which is perpendicular to the parental wall.

These rules were analyzed in 3D. It was found that there were several cell divisions in the embryo that were following the rule of minimal surface area that passes through the center of the cell. Jaipur these divisions are affected majorly by geometric properties of the cell.

But this rule was not followed by asymmetric divisions. But most of the cells started responding to the default division mode when there was an important regulatory pathway that got blocked in response to the phytohormone auxin.

Therefore, understanding this in a simple framework, do cell divisions follow or default division mode for dividing the cells symmetrically. The asymmetric divisions in the sense can be controlled and executed by disturbing this default mode.

There are several ways to manipulate this system it can be either done by alteration in the reference system of polarity, modification in the cytoskeleton or by converting the cell division process into the reference quality system. These critical problems in the polarity system are seen to be fully solved in the upcoming years.

1.3.4. Pattern Formation in the Early Embryo

With the increase in the number of cells in the embryo, the future organs are specified in the local regions of the embryo. In order to complete this process, there is a requirement of two elements that are essential for the completion of the process. Firstly, each and every pattern is needed to be specified. This can be achieved by installing a unique genetic program.

There has to sensible way two positions these organs in relation to each other at the same time. The two things required for the completion of the process of pattern formation is the specification and coordination. The cells are positioned using vectorially positional information which is the arrangement of the organs along the embryo axis that is top to bottom.

During the pattern formation in the early embryo, there is the identification of some molecules phytohormone auxin. Some other signaling molecules include the transport and proteins which have also been identified. The influx and efflux systems are used directionally transport this small molecule between the cells. This transportation system allows fluxes between the local aggregation and the cells.

The expression changes in a gene are triggered by the option locally which is the driving force for the developmental progression. The proper auxin response is the only factor that was found in an experiment in which all the virtually formation of important processes in the embryo depends on the proper auxin response.

Locally, there was inhibition of the auxin response in various domains of the embryo. Therefore, this hormone is responsible for the formation of a spatial pattern of organs in the plants.

Root formation is one of the most important outputs of auxin signaling. It mostly occurs at the lower end of the embryo. It is the place where there is an accumulation of a high level of auxin. Root formation involves the expression of a number of oxygen- dependent root- specific transcription factors. When it is expressed at any other place it is sufficient think for the establishment of an ectopic root.

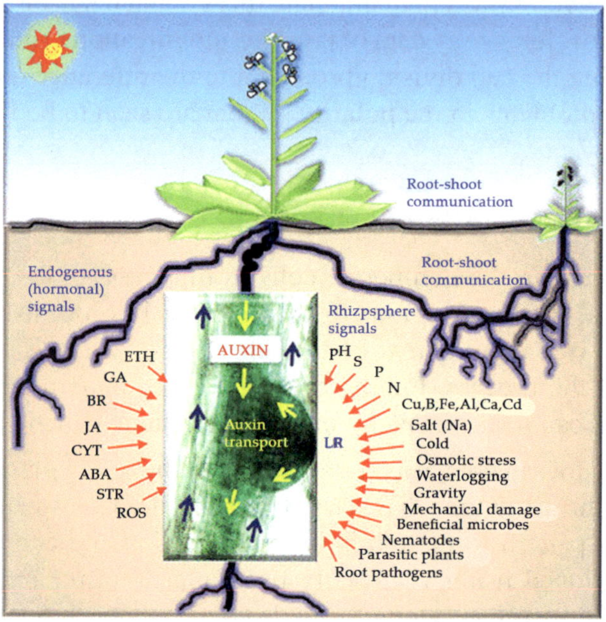

Figure 1.5: Root formation through auxin signaling

Source: Image by Flickr

There is an initiation of the shoot at the other end of the embryo. Later on, this area gets partitioned which is converted into cotyledons, shoot apical meristem and boundaries. These boundaries are helpful in separating the above two. The partitioning of the shoot region requires a number of genes. It has to be noted that for the cotyledon and the shoot apical meristem, a critical transcriptional regulator has to be known.

Therefore, the factors that are responsible for the organ pattern formation process are the vectorial information system which involves auxin transport and accumulation and the other factor is local decision making using key transcription factors. Further these are subdivided into specific functional organs.

The precursors of the major tissues within the embryonic organs are ground tissue, epidermis, and vascular tissue. This also includes the stem cells that work post embryonically for extending these tissues. The Identity of the tissue is at some point during the process of embryogenesis specified.

There is a process called tissue -specific gene expression reporters that can easily identify and visualize the tissue. However, it is very disappointing to say that there are very little information and mechanisms that can specify the identity of the tissue.

In order to establish the vascular and ground tissue, the response of auxin is critical. Where are different inputs are needed for the epidermis that relies totally on that? Among the three tissues which have been identified the only tissue that has a clear regulator which is sufficient for the specification is the epidermis.

Till the time there is no tissue-specific transcriptional regulator of ground tissue is known, other than the auxin response. In the epidermis, the factor that can define the tissue layers are the home domain transcription factors.

There is a loss of epidermal properties in the outer layer of the tissue due to the mutations present in these transcription factors. The differentiation between the sub- epidermal cell layer and epidermal cell types can be caused due to misexpression.

It is assumed that there are similar situations that exist for other tissues also. In the Arabidopsis embryo, there is no publication of any genome-wide map of gene expression at cellular resolution yet. Therefore, it is also unclear how large the differences among the types of tissues are. The fact not clear is which transcription factors make the different tissue types.

Several tissue -specific genes are still the subject of intense investigation, being known that several tissue-specific genes are isolated. It is assumed that several casual switches for different tissues will be identified in the coming years. In plant biology, the source of Nutrition is the plant embryo and it plays a vital role for fundamental questions in biology.

The mechanism of the ultimate cell identity switch that is from a somatic cell to an embryo cell is very wide in plants, however, this mechanism is not clearly understood.

The future of plant biology would be full of challenging situations one of them is to map the epigenetic trajectory of embryo initiation. Another challenging situation would be finding ways as to how the cell fate conversion is triggered and regulated.

Another key challenge in the process of pattern formation is to define how embryogenesis is different from other regenerative processes.

Other regenerative processes the body axis having shoot and root apical meristem is established de novo. However, there have been many studies regarding this which suggested that there may be significant overlap over these processes.

The process of plant embryos particularly in crucifers having no connection with the origin has provided the bulk of information on the tissue patterning and fundamental principles of morphogenesis.

That has been many improvements in the analysis of transcript me on a few single cells. In the future, it is assumed that with the advancement in live imaging, there will be many more countries that would give birth to the young plant.

Embryogenesis is basically the development in plants from the time of fertilization to the time of dormancy. The establishment of the basic body plan of the sporophyte completed during the process of embryogenesis. However, when the dormancy is broken, the plan has to be reiterated and elaborated again.

The major challenges in embryogenesis are:

- The establishment of the basic body plan of the embryo. Under this, there is the production of three tissue systems by the radial patterning and the establishment of the apical-basal (shoot-root) axis by the axial patterning.
- In order to make the postembryonic elaboration of the body structure, keeping aside the meristematic tissue (leaves, roots, flowers, etc.).
- Keeping an accessible food reserve for the embryo that is germinating until the time it becomes autotrophic.

The establishment of the basic body plan in embryogenesis is similar in all the angiosperms (Steeves and Sussex, 1989). The pattern elaboration has differences in the extent of shoot meristem development, the extent of endosperm development, cotyledon development, and the precision of cell division patterns. (Esau 1977; Johri et al. 1992).

After fertilization, polarity is established in the first cell division. The investigation of the polarity has been done by using a brown alga, as a model system (Belanger and Quatrano, 2000). The zygote so developed by these plants is independent of other tissues and is open for manipulation.

There are two types of cell formation in the initial cell division, one being the smaller cell and the other being the larger cell. The smaller cell that will ultimately form the rhizoid which is the root homolog and the larger cell anchors the rest of the plant that gives rise to the thallus, which is the main body of the sporophyte.

The fixation of the position of the rhizoid end of the apical-basal axis is done by the point of entry of the sperm. The axis stands perpendicular to the plane of the first division cell.At the rhizoid pole, there is an accumulation of F-actin (Kropf et al. 1999). However, this fixing of the axis can be altered through light or gravity. There will be a new position in the cell division (Alessa and Kropf, 1999).

The secretory vesicles are directed towards the rhizoid pole of the zygote, after the establishment of the apical-basal axis. The vesicles have all those materials which are essential for the outgrowth of rhizoid.

The orientation of the first plane of cell division gets easy by targeted secretion. The early development and the maintenance of rhizoid versus thallus fate totally depend on the information present in the cell walls (Brownlee and Berger,1995). In angiosperms, the information regarding the cell wall is also important (Scheres and Benfey, 1999).

Angiosperms also show the process of embryo stars with the basic body plan and asymmetrical cell division. The asymmetrical cell division gives rise to two types of cells that are a germinal cell and a basal cell. Do rise of the terminal cell rise to the embryo proper. The formation of the basal cell is made such that it is closest to the micropyle and finally it gives rise to the suspensor.

The interface between the embryo proper and the suspensor contains hypophysis. There is a rise of some of the root cells in other species. There is a division in the suspensor cell which forms a spherical organ and a filamentous organ that later gets be generated in the process of embryogenesis.

The suspensor Orient towards the surface of the embryo which is absorptive in nature towards the food source in both gymnosperms and angiosperms. Additionally, in angiosperms, the developing embryo also gets a nutrient supplement through the suspensor.

In Dicots, there is a need for a suspensor in the heart stage of the embryogenesis while culturing isolated embryos of scarlet runner bean switch or without suspensor (Yeung and Sussex, 1979). At this stage, a cultured embryo is assumed to survive which is cultured with a suspensor while an embryo cultured without a suspensor hardly survives at this stage.

Suspensor acts as a source of hormones. The young embryo in scarlet runner beans can survive without a suspensor and without culture provided they are given the growth hormone gibberellic acid (Cionini et al, 1976). The embryo proper and the suspensor develops some unique morphologies and it becomes useful to consider this as one of the key achievements of the embryogenesis process is the establishment of apical-basal polarity.

1.4. EMBRYOGENESIS IN HIGHER PLANTS: AN OVERVIEW

The formation of a progenitor cell of the endosperm and a single- celled zygote in higher plants is initiated with the double fertilization event with the sporophyte generation of higher plants. The embryogenesis is concerned with the formation of a zygote with a subsequent period of development.

This development can be described as a complex series in which the zygote undergoes the morphological and cellular changes.

These changes result in the formation of a developmentally fully mature embryo which is comprised of and embryonic axis with root poles, shoot poles and cotyledon. The embryonic axis has a high level of proteins and lipids which is considered as a good storage macromolecule.

The events occurred in the embryonic development makes the organization of the body of the plant that finally prepared the embryo for germination and dormancy. In the past century, there has been an intensive study on the embryogenesis in higher plants.

The embryonic development is characterized by the changes in the morphological and anatomical structure in the detailed studies using electron and light microscopy (Maheshwari, 1950; Wardlaw, 1955; Natesh and Rau, 1984; Raghavan, 1986).

The study of biosynthesis accumulation of lipids, proteins, macromolecular reserves, and starch is done with that of cellular differentiation. The post germinated seedlings are benefited as these ultimately serve as nutrients for the seeds (Jenner, 1982; Slack and Browse, 1984; Casey et al., 1986; Shotwell and Larkins, 1989).

The agricultural importance of these reserves can be detected by the level of attention on the reserves. Moreover, this reserve and the accumulation of these reserves help in further studies. Contrastingly there are lesser known facts about the processes that are under the genesis stages of the embryo, however much attention is devoted to this area.

The following are the processes involved in the formation of the embryo for which there is very little information. (Natesh and Rau, 1984; Dure, 1985; Raghavan, 1986; Crouch, 1987; Goldberg et al., 1989; Meinke, 1991a; De Jong et al., 1993; Lindsey and Topping, 1993).

1.5. MORPHOLOGICAL DESCRIPTION OF EMBRYOGENESIS

The early stages of embryogenesis are directly related to the events that occur during the process of gametophytic generation. It is necessary to understand the process of gametophyte generation to completely understand the morphological description of embryogenesis.

A Polar organization is there in the multicellular embryo sac embedded within the ovule. The polar organization is present along the micropylar-chalazal axis (Reiser and Fischer, 1993; Russell, 1993).

The micropylar pole of the ovule has egg cell and synergids to the closest of it, while at the opposite end of the embryo sac the antibody and cells are situated which is closest to the chalazal pole (Jensen, 1965; Schulz and Jensen, 1968a; Mansfield et al., 1990).

In the process of pollination, there is the penetration of the pollen tube into the ovule. This penetration is done through the micropyle which delivers a male gamete nucleus that enters the haploid nucleus and merges with the egg cell and finally produces the diploid embryo.

There may be a case that there is the production of an extra -embryonic triploid endosperm that happens when a second sperm nucleus is combined with the two polar nuclei. The two polar nuclei are located in the central cell which helps in the production of the triploid endosperm.

The nutrients reach the developing embryo through the endosperm. During the seed development, this nutrient is either absorbed or it persists there working as a strong tissue (Vijayaraghavan and Prabhakar, 1984; Lopes and Larkins, 1993, this issue).

The developing embryo is covered with the protective layer of maternal tissue of the ovule confines within this. Later when the embryo develops into the seed this protective layer becomes the seed coat that surrounds the developing embryo and endosperm. Embryogenesis in plants there are 3 overlapping phases that are divided conceptually.

The first phase of the embryogenesis in higher plants is of morphogenesis. During this first phase, the specification of the shoot and root apices are defined containing the polar axis of the plant body. In this phase, there is the formation of the embryonic tissue and the organ systems.

Embryo Maturation is the second phase of embryogenesis in higher plants. This phase is characterized by the accumulation of storage reserves.

In the final phase, the embryo enters into the period of developmental arrest as it prepares for desiccation and eventually gets desiccated. All the above mentioned stages are discussed below in detail and its emphasis can be applied to the morphogenetic phases.

1.5.1. Morphogenetic Stages

- Ovule

At the end of the micropylar (M) of the ovule, the synergids (SY) and the egg cells (EC) are located. While at the chalazal end (CH), the antipodal cells are located.

- Zygote
- One-celled embryo proper

The zygote here produces a larger basal cell and a smaller apical cell after going into the transverse cell division. Thereafter, the embryo proper is produced by the apical cell and the suspensor and hypophysis are developed by the basal cell.

- Two-celled embryo proper

The embryo proper is longitudinal in its first division and the transverse divisions divide the suspensor.

- Quadrant stage embryo

The four-celled embryo proper is produced with the help of two -celled embryo proper. The two-cell embryo proper is divided by another longitudinal division. It is divided longitudinally such that it is perpendicular to the plane of the previous division in the embryo. In this stage, the suspensor undergoes an additional transverse division.

- Octant stage embryo

Under this stage, there is the production of an eight-celled embryo proper through the four quadrants which are divided by transverse divisions. This division produces the transverse cell walls which later forms the indicated O' line. Under this, the basal portion of the suspensor is not shown.

- Dermatogen stage

Under this stage, there is the production of a sixteen-celled embryo proper which sets apart from protoderm. In order to form this sixteen-celled embryo proper, the cells of the octant stage embryo divides it by cleavages parallel to the surface.

- Early globular stage

Under this stage, in the cells of the protoderm, there is division perpendicular to the surface. There are additional longitudinal divisions in the interior cells of the embryo proper. In order to produce the hypophysis, the division of the topmost cell of the suspensor is done transversely.

- Mid-globular stage

The division of the cells of the hypophysis has been done longitudinally. While the cells present in the interior of the embryo proper are divided both longitudinally and transversely. The division of the protodermal cell is continued.

- Transition stage

As the apical pole of the embryo becomes broader, there is an indication given by the cell divisions which are parallel to the surface that the emergence of the cotyledon buttresses. The developing procambium becomes visible just like the elongated cells at the center of the embryo, ground meristem.

- Heart stage

In this stage, cotyledonary lobes continue to enlarge. It makes the change to bilateral symmetry obvious. Here, at this stage, O line is still recognizable.

- Linear cotyledon stage

This stage shows the morphological organization of the embryo. The cotyledons (c), the upper axis and the shoot apex (sa) ae there in the apical domain. The central domain consists of the bulk of the axis (ax)and the basal domain includes the root apex (ra). The developing vascular tissue forks just below the O boundary.

1.6. MAJOR PROCESSES IN EMBRYONIC DEVELOPMENT

In plants, there is a continuous occurrence of the formation of tissue and organ and it is considered as a striking characteristic in plants and is a repetitive process (Steves and Sussex, 1989). After this, the development of morphogenetic features such as the formation of roots stems leaves and other reproductive structures occur after the embryo is formed.

However, the establishment of the polar axis of a plant, features that define the organization of the plant body and primary tissue and organ systems are delineated during the embryonic development. The following is the discussion about the processes that are involved in making the embryo.

1.6.1. Embryonic Polarity

There is polarity in the plant body with the shoot root axis. It can be considered that a plant body is polarized by seeing the unfertilized egg cell. This unfertilized egg cell is present in an embryo sac which is organized in an asymmetric manner and ovule (Reiser and Fischer, 1993).

Many plants have the chalazal pole with a nucleus and much of the cytoplasm confined at the pole. While at micropylar end, there is the presence of a large vacuole.

However, do unfertilized egg cells which has the cell wall bounding the entire micropylar end is not completely surrounding the chalazal border (Jensen, 1965; Schulz and Jensen, 1968b; Mansfield et al., 1990; Russell, 1993, this issue). Following fertilization, the zygote is also characterized by an asymmetric organization.

The accentuation of the polar organization which is displayed in the Egg cell is done by redistribution of plastids, mitochondria and endoplasmic reticulum which occurs after fertilization (Jensen, 1968, Russell, 1993).

There are many other examples of polarization in the zygote that are described. For example, in Papaver Nudicaule and maize, after the fertilization, the nucleus and vacuole in the unfertilized eggs exchange their positions. After which the nucleus becomes oriented towards the chalazal pole (Olson and Cass, 1981; van Lammeren, 1981).

The polar organization of the egg is reflected by the asymmetry of the zygote. It is also considered that during the embryo sac development the embryonic polarity may have been established.

Just, for example, there are numerous plastid and cellular noise of mitochondria in the apical cell of cotton embryos, while the basal cell has many vacuoles (Jensen, 1968).

However, it has been observed that in some plants, the first division of the zygote is found oblique or symmetrical. Considering this it suggests that the organization of the zygote is more important than the plane of division.

What are those processes and signals that help in generating embryonic polarity? In the fruit fly drosophila, there are localized maternal signals that specify anterior- posterior and Dorsal- ventral axes of the embryo (Ingham, 1988; St Johnston and Nusslein-Volhard, 1992).

There are two observations that are inconsistent with the view that the polarity is fixed during the development of the embryo sac. This fact is suggested by the analysis of egg cell and zygote organization in plants.

As per the current knowledge and information, there are only one known maternal- effect mutations of higher plants that are the shrunken endosperm mutations of barley. This is considered to have affected the endosperm rather than its development (Felker et al., 1985).

Although, till now there is no identification of any other maternal effect mutants in higher plants, still it is possible that's such mutants good institute mutants of female sterile (Meinke, 1991b; Jürgens et al., 1991; Reiser and Fischer, 1993, this issue). There is only one mutation that can affect the asymmetric cleavage of zygote. This mutation can affect the zygotically acting gene, GNOM of Arabidopsis (Mayer et al., 1993).

Secondly, according to Zimmerman, 1993, somatic cells can mimic the maternal environment or there can be additional mechanisms that do not involve any maternal component that can generate polarity. This was suggested by Zimmerman, as in many plants somatic cells can be induced to undergo embryogenesis process.

The induction of cellular polarity in plant embryogenesis was studied using the fucoid algae zygotes and used as a model system (Quatrano, 1978; Kropf, 1992; Goodner and Quatrano, 1993). The variety of environmental stimuli that involves the microfilament networks and the cell wall results in the selection and fixation of a rhizoid thallus axis.

The egg cytoplasm being polar, the axis fixation of the alga zygote becomes a dynamic process in contrast to the zygotes in higher plants. The absence of the maternal factors in the egg cell which has the capacity to generate asymmetry in the absence of the clues about the maternal factors.

So, by focusing on the establishment of polarity may be obtained by studying plants. The cytoplasm of plants is organized after fertilization.

1.6.2. Pattern Formation: Establishing Embryonic Domains

The axis on which the elaboration of the plant body is dependent is defined by the polarity of the embryo. If the development of the embryo is seen conceptually, there are series of partitioning events that occur increasingly in more specialized regions.

There is the involvement of 3 spatial domains which is an early compartmentation step. These domains are along the longitudinal axis of the embryo (Jurgens et al., 1991; Mayer et al., 1991).

The cotyledons, upper hypocotyl, and the shoot apex constitute the apical domain. The majority of the hypocotyl constitutes in the central domain, and the basal domain consists mainly of the root. The studies of Arabidopsis mutants that are defective in the spatial organization of plants gives the initial evidence for these domains (Mayer et al., 1991).

The mutations which were identified in the saturation mutagenesis experiments and found to be four in nine complementation groups resulted in the deletion of the domains along the apical-basal axis. These were specifically for:

- The apical domain mutant, the shoot apical meristem, gurke and the cotyledons
- The central domain that lacks the mutant, in which the cotyledons are fused, fackel
- The basal deletion mutant, both the hypocotyl and root are lacking and monopteros
- The terminal mutant, both the apical and basal domains are lacking and gnom

According to these mutant phenotypes, it indicates that there are three distinct domains within the embryo. The radial pattern is affected by mutations in the five remaining complementation groups.

This is known to be a embryo shape and tissue organization. As per this fact, it can be assumed that the body organization of Arabidopsis requires 40 zygotically active genes which are considered as sufficient (Mayer et al., 1991). Since mutants are identified only at the seedling stage, it can be a probability that the defective mutant in the pattern formation can or not produce the seedling which is viable and which are unable to complete embryogenesis.

It is clear that the mutants at pattern formation were originally identified at the seedlings stage. Also, the observation as per the above discussion says that the consequences related to morphology is first observed at an early embryonic stage. This clears the fact that the domains are established early in embryogenesis (Mayer et al., 1991).

1.6.3. Establishment of Shoot and Root Apices and Embryonic Organs

The development of root and shoot in the embryo is together known as the embryonic axis and the polar axis of the embryo is defined with the specification of the shoot and root apical meristems. This formation of the embryonic root apex occurs with the help of plant cells of both the basal and apical cells of a two-celled embryo.

The Embryo proper region is incorporated with the hypophyseal region in order to give rise to part of the root cap and its initial cells and the ground meristem initial cells (Tykarska, 1979). The topmost region of the suspenser derives the hypophyseal region which is later combined with the embryo proper region.

The Apical cell helps in the contribution of the remainder of the root apex which includes the ground meristem and procambium. The heart stage is evident in the process wherein the root apical meristem organizes, however, delineation of the embryonic shoot apex is much more cryptic than this. The Globular stage embryo helps in identifying the epiphysis which is a shoot apical meristem in the very early stage.

Histological analysis indicates that the epiphysis is segregated early from the progenitor cells of the cotyledons. Also, it is said that the clonal analyses in cotton and maize suggest that the embryonic region gives rise to the shoot apical meristem, the epiphysis (Christianson, 1986; Poethig et al., 1986).

The epiphysis formation can also be used to study tissue differentiation in a way that it partitions in the apical domain that segregates cells with the potentia1 to become the shoot apical meristem with respect to the embryonic domains discussed previously. The analysis of 2S and 12S storage protein mRNA accumulation in oilseed rape embryos can help in inferring this study (Fernandez et al., 1991).

During embryogenesis, there is tight regulation of the storage protein gene expression. This marks cellular differentiation in an excellent way (Perez-Grau and Goldberg, 1989; Thomas, 1993).

A stage comes in the process of embryogenesis where the storage protein mRNAs are not detected in epiphyseal cells. However, they are found in the cotyledonary cells which indicates that the two regions are successfully segregated functionally within the apical domain.

Moreover, there is a distinct line of cell walls that helps in the separation of the lower boundary of the epiphysis and the adjacent cells with and without storage protein mRNA. This tissue type is transcended by the tissue types which later extends through the ground tissue and into the protoderm. Later, it coincides with the position of the O line.

The idea of the observation is unique and consistent that there is early partitioning of the progenitors of the shoot apex in the embryogenesis process. This is because there is the delineation of the protoderm due to the cell cleavage. The O line generated occurs just before the cell cleavage as the first transverse cell division of the embryo proper is responsible for this.

The cotyledons and the axis which is the morphological occurrence is formed during the transition from the globular stage to the heart stage. It is considered as the first visible sign of the formation of these two embryonic organ systems.

Later, in the apical region, there is an occurrence of another partitioning event that forms the cotyledons at the radially symmetrical globular stage. This indicates that at some specific site, there are a group of cells that are induced to do so.

According to a recent report, the localized cell divisions are directed by the involvement of the polar auxin transport (Liu et al., 1993). The embryos of the globular stage that are either the genetically found from plants that are defective in the polar auxin transport or treated with auxin transport inhibitors do not produce two cotyledonary lobes. Rather, it forms fused cotyledons which are basically a ring-shaped structure above shoot apices.

The globular stage is the only stage that produces the fused cotyledons. The inducement is particular to the stage. As, in the heart stage, auxin transport inhibitors applied to heart-shaped embryos do not induce the fused cotyledons.

As per the above findings, the auxin transport must be formed globular stage embryo so that it can indicate the site of cotyledon formation. It is also true that the auxin is perceived differently by different groups of cells in the apical region.

1.6.3. Tissue Differentiation

The primary embryonic tissues are organized in such a way within the embryo that it divides the embryo into different parts. The three main primary embryonic tissues found are termed as the protoderm, the ground tissue, and the procambium.

These tissues can be identified separately as histologically is the protoderm; progenitors of the ground tissue and procambium can generally be recognized during the transition from the globular stage to the heart stage of embryogenesis (Schulz and Jensen, 1968a; Tykarska, 1979; Mansfield and Briarty, 1990).

In the early heart stage embryos, the differentiation of these embryonic tissues is quite evident of their detailed structure, as plastids in cells of the ground meristem (a type of tissue) and protoderm become more differentiated than those in procambial cells, and ground meristem and procambial cells become more vacuolated than cells of the protoderm.

It has also been researched and suggested that the tissues can be differentiated before their detailed structure could be reviewed. For example, an early indication of protoderm differentiation is the division of cells occurring primarily by cleavages perpendicular to the embryo surface.

The Functional organization of these tissues can also be seen in early embryogenesis through their expressions of specific genes. The Accumulated mRNA that encodes a lipid transfer protein, EP2 also indicates that these cells are totally different from the interior cells of the embryo. The Protodermal cells in the globular stage carrot embryos specifically accumulate mRNA (Sterk et al., 1991).

Similarly, mRNA also encodes the Kunitz trypsin inhibitor, KTi3, which in turn accumulates in the ground meristem cells of soybean heart stage embryos. These can be easily found at the subset of cells at the micropylar end of the globular stage embryos that have been interpreted to be ground meristem initials (Perez-Grau and Goldberg, 1989).

Tissue differentiation starts at the globular stage and the cells that are destined to become specific embryonic tissues start to change at this stage only becoming unique and different from other tissues ultra-structurally.

The Formation and development of embryonic tissues, apical meristem, and organs are somehow dependent on Genetic approaches which gives an insight into the whole formation and development procedure. Embryonic mutants of Arabidopsis and maize have been described whose phenotypes suggest that morphogenetic processes have been altered (Meinke, 1986, 1991b; Clark and Sheridan, 1991).

Theses defective genes also alters cotyledon identity and the maturation program in such a way that storage proteins do not become prevalent in cotyledons of mutant embryos. Hence, the products derived through these defective genes appears to be the key embryonic regulators because of the alteration they cause.

Other than these, some embryos like Arabidopsis embryo have also been recognized which allows cellular differentiation but arrests morphological development at an early embryonic stage, as the maturation and desiccation stages depict about their expressions of genes (M.A.L. West, R. Yadegari, K.L. Matsudaira, J.L. Zimmerman, R.L. Fischer, R.B. Goldberg, and J.J. Harada, unpublished results).

These researches suggest the tissue differentiation in embryos can be released from morphogenesis. Other embryo mutations such as an Arabidopsis mutant that is a biotin auxotroph may represent disruptions in housekeeping functions, (Schneider et al., 1989).

1.7. CONCLUSION

The embryogenesis emerges the review that the process is seen as a hierarchy of partitioning events. Embryogenesis culminates in the production of organisms that are morphologically complex. The plant body is formed on an axis which is established as an asymmetric organization of the egg cell. This establishment is done by the polarity of the embryo.

The embryo whose compartmentation is done early sets off a domain. This domain helps in the establishment of the organization of the plant body. Early in the embryogenesis, the organization of other events such as embryonic tissue and organ systems and partition progenitors of the shoot and root apices is done. Also, at an early embryonic stage, the separating boundaries of the regions of the embryo coincide with cell cleavages.

However, in this instance, the difficulty is in distinguishing the roles of cell lineage and positional information. That distinguishing factor for determining the cell fate cannot be done as there is an establishment of consistent patterns of cleavages in cruciferous embryos that are made largely in the cell position.

Although, there is a provision regarding embryogenesis such as histological, morphological, and biochemical studies are there which have provided descriptive information about embryogenesis. However, there is a lack of mechanistic understanding of the processes that underlie embryonic development.

REFERENCES

1. Gilbert, S., n.d. Embryonic Development. [online] Ncbi.nlm.nih.gov. Available at: <https://www.ncbi.nlm.nih.gov/books/NBK10120/> [Accessed 26 May 2020].
2. Gupta, H., n.d. Embryo: Meaning, Development and Modes (With Diagrams) | Botany. [online] Biology Discussion. Available at: <https://www.biologydiscussion.com/embryology/embryo-meaning-development-and-modes-with-diagrams-botany/20796> [Accessed 26 May 2020].
3. Vries, S. and Weiner, D., n.d. Plant Embryogenesis. [eBook] Available at: <https://www.cell.com/current-biology/pdf/S0960-9822(17)30562-6.pdf> [Accessed 26 May 2020].
4. West, M. and Harada, J., n.d. Embryogenesis in Higher Plants: An Overview. [eBook] Available at: <https://www.ncbi.nlm.nih.gov/pmc/articles/PMC160368/pdf/051361.pdf> [Accessed 26 May 2020].
5. Alessa and Kropf 1999
6. Belanger and Quatrano 2000
7. Brownlee and Berger,1995
8. Casey et al., 1986
9. Cionini et al. 1976
10. Crouch, 1987
11. De Jong et al., 1993
12. Dure, 1985
13. Esau 1977; Johri et al. 1992
14. Goldberg et al., 1989
15. Goodner and Quatrano, 1993
16. Ingham, 1988
17. Jenner, 1982
18. Jensen, 1965
19. Jensen, 1968, Russell, 1993
20. Jürgens et al., 1991
21. Kropf et al. 1999
22. Kropf, 1992
23. Lindsey and Topping, 1993

24. Mansfield et al., 1990
25. Mayer et al., 1993
26. Meinke, 1991b
27. Natesh and Rau, 1984
28. Olson and Cass, 1981; van Lammeren, 1981
29. Quatrano, 1978
30. Raghavan, 1986
31. Reiser and Fischer, 1993
32. Russell, 1993
33. Scheres and Benfey, 1999
34. Schnarf (1929), Johansen (1945) and Maheshwari (1950)
35. Schulz and Jensen, 1968b
36. Shotwell and Larkins, 1989
37. Slack and Browse, 1984
38. St Johnston and Nusslein-Volhard, 1992
39. Steeves and Sussex 1989
40. Yeung and Sussex 1979

Chapter 2

Plant Reproduction

CONTENTS

This chapter is about the reproduction process that takes place in the plants in order to produce the young ones or new plant from the parent plant. This chapter gives a deep knowledge of the process of reproduction in plants.

As the reproduction process in different plants takes place through different methods. These methods are vegetative, asexual, and sexual. All these methods have different sub -methods which are being used by the plant for the production of a new plant.

Similarly, the vegetative method has different sub -methods like cutting, budding, layering, separating, and grafting which are being used for producing the new plant which is similar to their parent plant. The sexual reproduction process takes place through pollination.

Pollination is a process known as the sexual process which is used by the plants for reproduction. In this method, there is an involvement of male and female gametes that fuses together to form a new plant through an embryo. There are typically two types of pollination methods like self-pollination and cross-pollination which is being explained in detail.

2.1. INTRODUCTION

Reproduction is a process that takes place in order to produce the new offspring in the plants. Therefore, the reproduction process is required for the generation of new individuals. Plants that are present on the land are considered to have asexual reproduction process, where the offspring are found to be genetically identical to their parents or sexual by creating a genetic variation.

The life cycle of the plants basically involves two alternating generations: sporophytes that are making the spores and the gametophyte that is producing the make and the female gametes. During the evolution in the land plant, the life cycle has been altered from the gametophyte generation.

It is observed that the structure that is making the gametes and spores, and the way through which the reproductive cells are being exchanged and dispersed. It also becomes altered, as the plants become to adapt the living in the drier environments.

Flowers are representing the reproductive organ of the flowering plants and considered to be very important for the identification as it is typically providing the characters that are steadily being expressed within a taxon.

This is because the floral characteristics are basically under strong genetic control and it is not affected by the changes in the environments. There are certain floral characters that may remain the same throughout the genus or a family, while other characteristics have become more variable and are being used to differentiate the species.

For example, the floral symmetry, type of placentation, ovary position and kind of fruit that are usually used to differentiate the genera or families, while the shape, size of the floral parts and the petal color are being used most commonly for distinguishing the species.

Flowers basically arise from the apical portion of a stem in the region that is known as a receptacle. The receptacle may be borne directly on the main stem axis or rachis (sessile) or a stem called a pedicel or on a slender stalk.

They are usually found to consist of the four whorls of parts that develop in the flowing series. The series starts from the outer whorl to inner: sepals, petals, stamens, and carpels. These parts of whorls are developing in a discrete cycle or found to be more or less spirally and continuously arranges.

If all the four parts of the whorls are present, the flower is then said to be complete. If anyone of the whorls is found to be missing, the flower is said to be incomplete. The symmetric of the flower is usually defined as a radial (actinomorphic) or bilateral (zygomorphic). There are few groups of the flowers that are termed as irregular (asymmetric) when the sepals or the petals are found to be dissimilar the form or orientation.

There is a possibility that the flower on a given plant may occur aggregated or singly into a cluster, which is called inflorescences. The supporting stem or the stock of the inflorescence or may be of an individual flower is known as a peduncle.

There are basically two main types of inflorescences which are: determinate, in which the oldest flower is found to be in the center or above others. Second is indeterminate, in this, the terminal flower is said to lack and the flowers are usually found to be axillary from the leaves or bracts and mature basically from the base to the top (for example, panicle, corymb, raceme, spike, spikelet and spadix).

There are certain inflorescences like head or capitulum, catkin that can be either determinate or indeterminate, this totally depends on the plant group. Inflorescences may undertake different shapes and they may also not show the branching.

The identification of the type of inflorescence is being made by simply determining the oldest flower position and nature and the extent of branching, the presence or absence of pedicels and the general shape of the structure.

2.2. MORPHOLOGY OF A FLOWER

Flowers are considered as the reproductive organs of the plant and are the knowledge of various parts of a typical flower which is said to be important for understanding the plant sexual reproduction. A flower basically consists of different floral whorls. Each of these whorls is with different functions.

The outermost of the whorl is called a calyx and consists of a sepal. Sepals are found to be green in color and they enclose and protect the developing bud. The whorl which is next to the calyx is the corolla, which is consisting of petals and usually helps to attract the pollinators.

Some of the species like tulips, in them the sepals and petals look very much alike and they also act together in order to provide the color which helps in attracting the pollinators. Together, the calyx and corolla make up the perianth. The whorl which is next to the corolla is androecium, which is consisting of male organs which are called stamens.

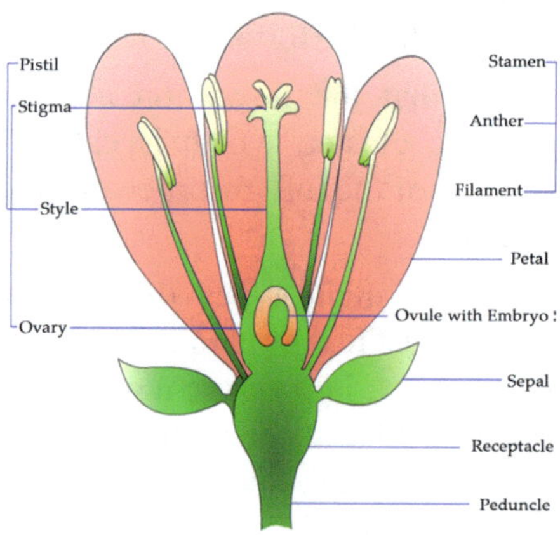

Figure 2.1: Structure of a flower.

Source: Image by Glenbrae School Room 10

Each of these stamens are typically found to consist of a slender stalk or the filament that attaches to the flower at its base and is consider to carry on its free, upper end, a structure that is called an anther which contains the pollen.

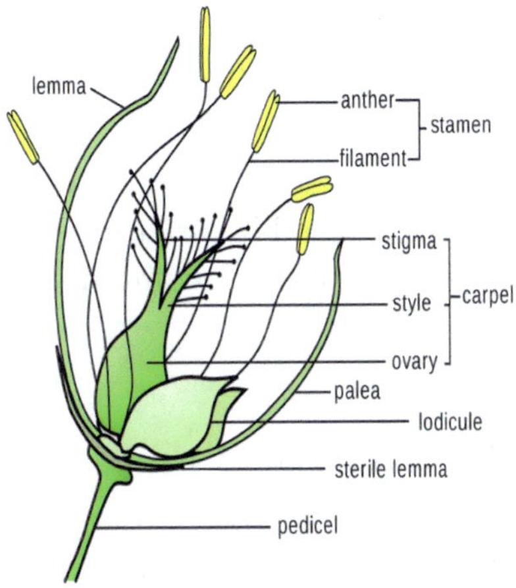

Figure 2.2: An image representing female and male organ systems.

Source: Image by Wikipedia

The innermost whorl of the flower, which is called as gynoecium, consists of a female organ which is known as carpels. Each of these carpels is consisting of a basal ovary which is containing the ovules, which is a slender column-shaped structure, the style and at the end of the style the stigma, which is used to receive the pollen grains.

Additionally, the whorls like the epicalyx are consisting of a bract which basically occurs outside the calyx and they may also be present in some of the flowers like cotton.

The floral morphology which is discussed above is considered very typical, but in the real world, it is considered that the logical pattern of the flower parts is not so clear. In a typical flower, there are some parts that may also appear similar or some of the parts may also be missing or some parts or groups of the whorls may get merged.

Therefore, the most common instance of similarity among the parts is the resemblance that is between the sepals and which might have been already mentioned as occurring in tulips. Similarly, the brightly colored leaves and bracts which are surrounding the flower may also found to be confusing with the petals, for example, Bougainvillea.

There are many species that have evolutionarily lost some of the parts of the flowers. The most obvious situation is when a plant or a species are having different male and female flowers. In this case, the flowers might have lost one of their sexual function, allowing them to get specialize in the other.

Cohesion and fusion are said to be common in both within and among the flowers. The petals may get fused in order to make a tube, as in the case of a petunia flower. Flowers may also get combine to form the inflorescence as in Brassicas.

2.3. MODES OF REPRODUCTION

There are various modes of reproduction through which plants reproduce: Vegetative, Asexual, Sexual. In the Asexual and vegetative mode of reproduction, it is considered that the offspring are being produced from the vegetative unit which is formed by a parent without having any fusion of gametes or sex cells. In this process, a single parent is being involved and offspring are generally found to be identical to the parents.

2.3.1. Vegetative Reproduction

- **Vegetative Reproduction:** It is the process that involves the formation of new plantlets from the vegetative cell, buds, or organs of the plants. In this type of reproduction, the vegetative part of the plant like Stem, Root, shoot, or bud are considered to get detached from the parent body and then it grows into an individual or independent plant.

It is found to be similar to the asexual reproduction in which it requires only the mitotic division, there is no gamete fusion involvement, and the newly formed plants is considered to be genetically clone of the parent plant.

- **Fragmentation:** Like in case of filamentous algae, there is an accidental break down of the filaments into many fragments, where each of the fragment is considered to have at least one cell, and it may also give rise to the new filament of the algae by using the process of cell division. For example, Spirogyra.
- **Fission:** the process of fission basically occurs in the unicellular organisms like yeast and bacteria, where the content of the parent

cell is said to divide into 2, 4, or 8 daughter cells and according to the fission it is known as a binary or multiple fission. Each of these newly formed daughter cells is considered to grow into a new organism.

- **Budding:** Budding is a process that also occurs in the unicellular plants. In this process, a bud- like outgrowth is Heredity formed on one side of the parent cell and is also soon get separated and grows into a new individual. For example, Yeast.

2.3.2. Asexual Reproduction

Asexual reproduction takes place by the asexual spores which can be flagellate or can be non-flagellate.

- **Spore Formation:** Spore formation takes place in the lower plants, including pteridophytes or bryophytes. There are special reproductive units that can develop asexually on the body of the parent. These are called spores.

These spores are microscopic and are being covered by the protective wall. When the spores reach a suitable environment then they are said to develop into a new plant body. For example, in the bread molds, moss and fern.

In some of the higher plants like maize, gymnosperms and pea are said to reproduce asexually which is always heterosporous. The spores in this case, are formed after the process of meiosis. The small male spores are called microspores which give rise to the male gametophytes.

The large female spores are called megaspores and they are said to give rise to the female gametophytes.

2.3.3. Sexual Reproduction

- **Ovule Formation:** All the plants are considered to have a life cycle that consists of two distinct forms that basically differ in size and in the number of chromosomes per cell. In the flowering plants, there are the large and familiar form that consists of roots, shoots, leaves and reproductive structures which is diploid (like fruits and flowers) and is known as the sporophyte.

Sporophytes are producing haploid microscopic gametophytes that are said to be dependent on the tissues which are produced by the flower. The reproductive cycle of a flowering plant is basically regular, usually seasonal, cycling back and forth from the sporophyte to the gametophyte.

The flowers are said to produce two kinds of gametophytes, namely female and male. The female gametophyte arises from the cell with the ovule, which is a small structure within the ovary of the flower. The ovary is considered as a large structure present within the flower and is said to contain and protect many ovules.

Flowering plants are considered as unique in which their ovules are found to be entirely closed in the ovary. The ovary is a part of a larger structure that is known as a carpel, which basically consists of the stigma, ovary, and style.

Each of the ovules is found to be attached to the ovary tissues by a stalk called the funicle. The point at which the funicle attached to the ovary is known as the placenta. As the flower is developing from the bud, therefore the cell within an ovule is called as the arche spores that enlarges in order to form an Embryo Sac Mother Cell (EMS).

EMC is divided by the process of meiosis in order to produce four megaspores. In this process, the number of chromosomes that is reduced from the two sets in the EMC to one of the sets in the megaspores, thus making the megaspores haploid.

Three of the four megaspores are basically found to degenerate and disappear, while the fourth is observed to divided mitotically three times in order to produce the eight haploid cells. These cells together constitute the female gametophyte, which is called the embryo sac.

The eighth embryo sac cells, often differentiate into two synergids, that is three antipodal cells, two fused endosperm nuclei, and at last an egg cell. The mature embryo sac is observed to be situated at the outer opening (known as micropyle) of the ovule, which is found to be ready for receiving the sperm cells which is delivered by the male gametophyte.

- **Pollen:** The mature pollen is the male gametophyte. Pollen is being produced in the anther, which is basically attached at the distal end of the filaments. The filament together with the anther is said to constitute the stamen, which is the male sex organ in flowers.

Flowers are usually producing many stamens inside of the petals. When the flower gets mature, then the cell in the anther is divided mitotically in order to produce the pollen mother cells (PMC).

The Pollen mother cells are further divided by the meiosis to produce the haploid microspores in the groups of four which are called tetrads. The micropores are found within a single layer of the cell which is called the tapetum. This layer provides nutrition to the developing pollen grains.

Each of these microspores develops into a hard, opaque outer layer which is called the exine. This layer is constructed from the lipoprotein which is called sporopollenin. It is considered that the exine contains the characteristic pores, ridges or projections that can also be used for the identification of the species, even in the fossil pollen.

The microspores are divided mitotically once or twice in order to produce two or three haploid nuclei inside the mature pollen grains. Where, two of the nuclei are said to function as a sperm nucleus that can finally fuse with the egg and endosperm nuclei of the embryo sac. Which in turn produces, an embryo and the endosperm, respectively.

For the process of sexual fusion to take place, the pollen grains are needed to be transported to the stigma, which acts as a receptive platform present on the top of the style, which is an elongated extension on the top of the carpel.

The moist surface or chemicals present on the top of the carpel helps the pollen grain to germinate. Germination is basically the growth of a tube from the surface of the pollen grains.

The tube is considered as a sheath of the pectin, inside of which there is the presence of a solution of water, solutes and two or three nuclei, which may also lack any cell wall.

The proper growth of the pollen tube often requires an aqueous solution of an appropriate concentration of the solute, as well as the nutrients such as boron, which may be considered as an aid in the synthesis of the protein.

At the apex of the tube, there are active ribosomes and also endoplasmic reticulum which is a type of organelles found to be involved in the process of protein synthesis. Pectinase and a glucanase which probably helps in maintaining the flexibility of the growing tube and also aids in the penetration.

The pollen tube apex is also considered to release the ribonucleic acid (RNA) and the ribosomes, in the tissues of the style. The tube generally grows to eventually reach the ovary. It is the space where it travels along with the intercellular space until it reaches the placenta.

Through the chemical recognition, the pollen tubes change its direction of the growth and it penetrates through the placenta to the ovule. The tube then reaches the embryo sac which is lying close to the micropyle and then the sexual fertilization process takes place.

- **Double Fertilization:** Fertilization process in the flowering plants is considered as unique among all the known organisms. In this process, instead of one cell, two cell gets fertilized and then this process is known as double fertilization.

One sperm nucleus present in the pollen tube fuses with the egg cell present in the embryo sac, whereas another sperm nucleus fuses with the diploid endosperm nucleus. The fertilized egg cell that is present in the zygote further develops into the diploid embryo of the sporophyte.

The fertilized endosperm nucleus later develops into the triploid endosperm which is considered as a nutritive tissue that can sustain the embryo and seedling too. Other known plant groups are considered to exhibit the double fertilization process in the Gnetales present in the genus *Ephedra*, which is basically a nonflowering seed plant.

Therefore, in this case the process of second fertilization product often degenerates and does not even develop into the endosperm.

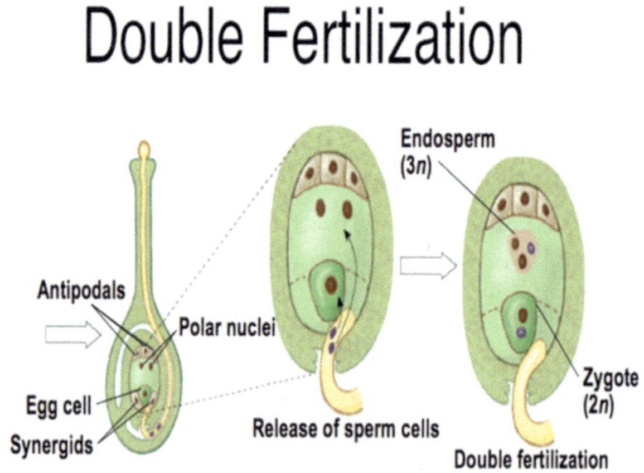

Figure 2.3: Representation of double fertilization process.

Source: Image by Toppr

Double fertilization begins when the pollen tube grows into one of the two synergid cells present in the embryo sac. This process is considered a result of chemical attraction that occurs due to the presence of calcium.

After the penetration of the synergids, the apex of the pollen tube breaks and opens, which releases the two sperm nuclei and with that, there is a release of other contents into the synergids. As the synergids degenerate, it is believed that it envelops the egg and endosperm cells by simply holding the two sperm nuclei which are close and the other expelled contents present in the pollen tube.

The egg cells, after then opens and engulfs the sperm cells. The egg cell membrane breaks and after then it allows the nucleus to move near the egg nucleus. The nuclear then envelopes the disintegrate and then combines the two nuclei in order to form the single diploid nucleus of the zygote.

Later, the other sperm cell fuses with the two endosperm nuclei. This helps in forming a single triploid cell, the endosperm cells, which gets divided by the help of mitosis into the endosperm tissues.

- **The Seed:** The seed is considered as a mature and fertilized ovule. After the process of fertilization, the haploid cells of the embryo sac get disintegrated. The diploid cells of the ovule which is derived maternally develop into the hard, water- resistant outer covering of the seed which is called a testa or a seed coat.

The diploid zygotes are further observed to develop into the embryo. Whereas, the triploid endosperm cells multiply and also provide nutrition. The testa is usually seen to show the scars which are called the hilum, where the ovule is seen to be originally attached to the funicle.

Whereas, in some of the seeds, it is being observed that seeds are ridge along the testa which is called the raphe which often shows exactly where the funicle originally pressed against the ovule. The micropyle of the ovary is usually seen to survive as a small pore present in the seed coat that basically allows the passage of the water during the process of germination of the seed.

In some of the species, it is observed that the funicle develops into a larger structure present on the seed which is called an aril. Aril is seen as brightly colored, contains sugar and juicy, that is consumed especially by the animals that may also disperse the seeds which can be in the form of nutmeg, oxalis, castor beans and arrowroot. Aril is considered to be distinct from the fruits, which basically forms from the ovary itself.

The embryo is said to contain the cotyledons, hypocotyl, and epicotyl. The cotyledons look like a small leave and are usually considered as the first photosynthetic organ of the plant. Epicotyl is the portion of the embryo which is basically present above the cotyledons and the portion that is below is called a hypocotyl.

The epicotyl is basically an apical meristem that produces the shoot of the growing plant and is also considered as the first true leaves after the germination. The hypocotyl is found to develop in the roots.

Even the tip of the hypocotyl, the radicle is considered as the first indication for the germination, as it often breaks out of the seed. The flowering plants are being classified as a monocotyledon or the dicotyledon which is basically based on the number of cotyledons that are produced in the embryo.

The common monocotyledons basically include the grasses, lilies, orchids, irises, and sedges. The common cotyledons include the roses, snapdragon, sunflowers and all the non-coniferous trees.

The endosperm may also get consumed by the embryo, as in the case of many legumes, which basically uses the cotyledons as a source of food during the process of germination. In other species, it is found that the endosperm persists until the process of germination when it is being used as a food reserve.

In the grains, such as wheat and corn, their outer layer of the endosperm is said to consist of the thick wall cells called aleurone, which is basically a protein.

- **The Fruit:** the fruit of the flowering plant is considered as a mature ovary. As the seeds get mature, the surrounding ovary wall is said to form a protective structure that may also aid in the dispersal. The surrounding ovary tissue is known as the pericarp and it basically consists of three layers.

These layers are the endoocarp, mesocarp, and exocarp from inside to outside. The exocarp is usually known to be skin- like and tough. The mesocarp is seen to be often thick, sweet, and succulent.

The endocarp which is basically surrounding the seeds is found to be hard and stony. The fruits are simply termed by the single ripened ovary, which is present in the single flower- like apricot, blackberry, and strawberries.

Multiple fruits are found to be in a cluster of many ripened ovaries that are present on the separate flowers which are basically growing together in the same inflorescence like fig, pineapple, and mulberry.

A simple fruit can be found as dry or fleshy. A fleshy simple fruit that is being classified as a berry include- tomato, papaya and grape, hesperidium like orange, pome like apple, pepo like pumpkin, cucumber and watermelon, or a drupe like apricot.

It is observed that the dry simple fruit will have a dry pericarp at the time of maturity. They may or they may not dehisce or split along a seam in order to release the seeds. The dehiscent dry fruit is being classified as a capsule like poppy, lily, silique or a silicle like mustard, follicle like larkspur, milkweed, and columbine, or as a legume or pod like bean and pea.

There is an indehiscent dry fruit that does not split for releasing the seeds which are classified as schizocarp: like funnel, carrot and celery, nut: like chestnut, hazelnut and acorn, grain or caryopsis like grasses such as rice, wheat and barley, winged samara like elm, maple and ash, utricle like duckweed family, or an achene like buttercup, sycamore and sunflower.

There are some of the fruiting bodies that are considered to contain a non -ovary tissue and are also sometimes called as pseudocarps. The sweet flesh of an apple and pears is being composed not of the pericarp but of the receptacle, or the upper portion of the flowering shoot in order to which the petals and other floral organs are being attached.

Fruiting bodies are considered to contain all kinds of functions that are in order to protect and disperse the seeds they contain. Protection can be chemical in terms of repellents of seed predators or can be physical in terms of hard coverings.

Sweety and fleshy fruits are considered as an attractive food for the mammals and birds that are said to consume seeds along with the fruits and often pass the seeds to get interacted with their fecal matter, which will often act as a fertilizer.

Dry fruits are usually considered to adapt for the wind for dispersal of the seeds, for example, the assistance of the wings like structures or a fluffy pappus that will often provide buoyancy. The diversity of fruiting bodies is considered to reflect in the part the diversity of dispersal agents in the environment, which will often select for the different fruit size, chemistry, and shape.

- **Pollination and Pollinators:** Pollination is a process of movement of the pollen from the stamens to the stigma, where the process of germination and growth of the pollen tube occurs.

Approximately 96 percent of all the flowering plant species are considered to be hermaphroditic which possess both the sexual functions within a plant and usually within every flower, and thus an individual can get pollinated on its own or by the pollen from another individual.

Seeds that are producing through the process of self-pollination and is often considered to be inferior in the growth, fecundity, and survival of seeds which is produced through the outcross pollination.

In most of the species, it is found that there is strong natural selection in order to minimize the proportion of the outcrossed seed.

The flowering plants are usually considered to be among the seed plants in their superlative exploitation of the animals as the agents of the outcross pollination. The outcross pollination efficiency of the insects, mammals, and birds (primarily bats) might have also contributed to both the diversity and abundance of flowering plants.

The abundance of flowering plants may have increased because of the less wastage of resources and energy on the unsuccessful ovules and pollens.

There might be an increase in the diversity that may be because of two reasons. First, the insects undoubtedly have been selected for a wide variety of floral forms that basically provide different rewards like nectar and pollen and are also appear to be attractive like size, shape and color, and scent like skunky and sweet that is also in different ways to different pollens.

The second is the faithfulness of the pollinators towards the familiar flowers that might have reduced the hybridization and speeded the evolutionary divergence and production of the new species.

Even though the flowering plants are first to appear after the discovery of all the major groups of insects. Flowering plants are believed to cause the evolution of many new species present within these groups.

There are some new insect groups like butterflies and bees, those that have originated after the flowering of the plants. Their members have been observed to develop their mouthpart structure and behavior that is specialized for pollination.

There are some extreme cases in which a plant is completely dependent on one insect species for the pollination, and the insect on the other hand is found to be completely dependent on the one plant species for their food.

Such an association is said to occur rarely but it is being observed in fig trees/some fig wasps, yuccas/yucca moths, and senita cacti/senita moth, which is also well documented. In all these three insects, the female is considered to lay eggs in the flowers and their young ones are seen to hatch later to feed on the mature fruits and its contents.

Even the female also ensures that the fruit develops by gathering the pollen from another plant and hence transporting it to the stigma of the flowers in order to hold their eggs. Plants are considered to benefit greatly in the outcrossed seed production.

At a very low cost or small cost, it is believed that some consumed seeds and fruits, and the insects get benefit greatly from the food supply for the developing larvae at a very small cost of transporting the pollen at the short distance between the plants.

Pollinating agents, whether they are biotic or abiotic, are considered to have an exerted strong selection on all the aspects of the flowers. This results in the evolution of the tremendous floral diversity. This diversity is believed to be distilled into a small number of characteristic pollination syndrome.

Pollination by the beetles requires white color, a shallow, bowl -shaped, and a strong fruity scent flower. Bees are selecting the yellow or blue or purple colorings, a landing platform with the color patterns that will help them to guide the bee towards the nectar, bilateral symmetry and with a sweet scent.

Butterflies are selecting for many colors, but they are selecting yellow, a corolla tube with the nectar at the base and in the absence of any type of scent. Whereas, the moths are selecting in contrast for nocturnally opening flowers with the strong scent and are drab and white in color, and a tube with the nectar at the base.

Bats are considered to select for the nocturnally opening flowers. But these flowers are with a strong musky scent and copious nectar, which is positioned well outside the foliage for easy access and drab or white color.

Hummingbirds are selecting red or orange flowers those are basically with no scent, copious nectar production and corolla tube with the nectar present at the base. There are other pollinating birds that do not hover while the feeding select for the strong perches and flowers are often found to be capable of containing copious nectar in cups, tubes, or funnels.

Wind act as a pollinating agent which often selects the lack of color, scent, small corolla, buoyant pollen, a large stigmatic surface area, nectar, hanging stamens and usually erect styles and limp.

Additionally, there is a great floral diversity present within any of these syndromes and considered to arise from the diverse evolutionary histories of the member plant species.

- **Selfing and Outcrossing:** Most of the flowering plant species are reproducing primarily by the process of outcrossing, including the great majority of trees, herbs, and shrubs. An adaptation that is helping in the prevention of self-fertilization basically includes the self-incompatibility like blocking the self-pollen and genetic recognition, and dioecy which includes separate female and male individuals.

Adaptation is the process that helps in reducing the chances of self-pollination in the hermaphrodites which include separation of the stigma and anther in the space (called as herkogamy) or time (dichogamy). There are many species, those are self-incompatible and there is a spatiotemporal separation of the sex organs.

Therefore, the ability to produce the seeds by the process of Selfing is considered as advantageous in the situation where there is outcrossing pollination is found to be difficult or impossible.

These further include the harsh environment where the pollinators are expected to be rare or unpredictable, and are regularly disturbed ground where the survivors often end up isolated from each other.

Selfing is considered to be less energy requiring than the outcrossing because the selfers have a chance of pollination with the assistance from the animals and not produce the large, attractive flowers that contain abundant nectar and pollen rewards.

Most of the primarily Selfing species are observed to be small annuals in the variable or disturbed habitats with the small, drab flowers. Selfers are mostly the desert annuals and the roadside weeds.

The evolutionary transitions are from the outcrossing to the near-complete Selfing that might have occurred many times in the flowering plants. Outcrossing and the Selfing species are considered to differ in their evolutionary potential.

Out-crossers are considered to be generally more genetically diverse and are also considered to produce the lineages that are said to persist over a long period of evolutionary time. During this time many new species are being formed.

Selfers are however found to be less genetically diverse and also tend to accumulate the harmful mutations. They are typically found to go extinct before they gather an opportunity to evolve the new species.

2.4. METHODS OF VEGETATIVE PLANT REPRODUCTION

The most widely used vegetative or asexual plant reproduction method is the propagation by cutting. There are other methods that include propagation by layering, division, and separation, budding and grafting, and tissue culture.

2.4.1. Cuttings

Cutting is considered as any plant part that severed from the parent plant. Cutting often used in the asexual reproduction, including those from the roots, stems, and leaves. For the successful reproduction, it is required that the severed plant parts must be developed for those parts that make it to complete the plant.

A stem cutting system is used to develop a new root system. The root cutting system should also develop a new shoot system. Cutting from the leaves must produce both the new root and shoot system. The stem cutting is being taken from the stems of a plant.

For the asexual reproduction method, a portion of a stem is inserted which includes node into a rooting medium. After several weeks, there is the formation of the roots, which occurs along the portion of the stem that takes place in the rooting medium.

The use of a stem cutting is considered as a successful method that is used for reproducing many types of ornamental plants, and several horticultural crops like grapes, figs, sweet potatoes, and citrus.

Stem cutting is being classified according to their stages of growth like herbaceous, softwood, semi- hardwood and hardwood. There is the use of four types of stems cutting for asexually reproducing plants. These types include tip cuttings, single-eye cuttings, cane cuttings, and medical cuttings.

Figure 2.4: An image representing the correct method of cutting.

Source: Image by Pinterest

Modified stems cutting those are taken from the tubers and from the rhizomes that are important for the process of reproduction of white potatoes, cassava, and certain grasses. Each of the cutting often requires a thickened area for the root and shoot development to occur.

Root cutting often consists of sections of thickened root that is cut into 2 to 3-inch sections and is being placed in a rooting medium. It is considered that they do not have any leaves or stem attached. Plants that are reproducing asexually from the root cuttings are those that are naturally produced sprouts or suckers from the root system.

Pecan, pear, apple, sweet potato, blackberries, and horseradish are few of the examples of fruits, nuts and vegetables that are being reproduced by using root cuttings. Leaf cuttings basically consist of a leaf blade and petiole, it is believed that it will contain either only a leaf blade or a section of a leaf blade.

It is believed that they do not include any stem tissue or a lateral bud. There is a use of leaf cuttings in order to have asexually reproduced herbaceous plants that contain a basal rosette arrangement of the leaves. This method is considered as a common method that is used for reproducing many of the ornamental and greenhouse plants.

2.4.2. Layering

Layering basically involves the forcing a part of a plant in order to form the roots by remaining attached to the parent plants. Air layering and ground layering are basically two common layering techniques. Air layering involves the forcing of roots to form on a stem, basically outside the soil.

While growing, a girdle cut or a cleft cut is made on the above -ground section of a stem and then covering is done with a moisture- holding media. Root formation basically occurs on the plant stem, specifically in the area of the cut. Whereas, the stem remains attached to the parent plant.

Ground layering involves the extending part of a plant into the ground by covering it with the soil and with that allowing it to the root. In order to encourage the root formation on the stem, a cut is being made on the section of the stem that is covered with the soil.

Ground layering methods include simple layering, mound layering and trench layering. After the roots are developed on the stem in the area of the cut, the removal of the rooted stem takes place from the parent plant in order to form a new plant.

Layering is basically a simple method of asexual plant reproduction and is considered to be effective for the propagation of some of the plants that are found to be difficult to reproduce by other methods. It is also considered as a cost-effective method used for the plant propagation that allows producing a large plant in a short period.

2.4.3. Separation and Division

There are many plants that are having specialized vegetative structures. Corms, rhizomes, bulbs, stolons, and tubers are few of the examples of specialized vegetative structures. These specialized vegetative structures are considered to contain a vegetative bud that is found to be capable of producing the new shoots.

The newly developed shoots are considered to eventually form roots and develop into the entire plants. The division and separation are two methods that are being used for producing new plants from their specialized vegetative structures.

Separation involves the removal of new plants that are formed from the specialized stems and even separating them from the parent plants. The division is considered as the technique that is used for cutting the specialized plant structures into the sections and even forcing each section to grow into a new plant.

2.4.4. Grafting and Budding

Grafting is basically the oldest method that is used as asexual reproduction. This process is being practiced over two thousand years. Grafting is a process that is basically consisting joining of a hardwood scion from a plant to the rootstock of another hardwood plant.

In most cases, the scion and rootstock wood are being used are of the same plant. However, grafting of the wood between the two closely related plants can be also possible. An example of the former that is using grafting of the paper shell pecan wood onto a native pecan rootstock.

A good example of the latter is the grafting of the plum scion to peach as the rootstock. The size of scion and the rootstock is being determined by using the grafting method. Whip and tongue, wedge, approach grafting and splice are basically used when both the scion and the rootstock are considered to be similar in size.

Bark inlay, cleft, notch, and the side grafting are commonly used when the scion is observed to be smaller in the diameter than the rootstock. Side grafting is basically used for adding the branches to bare the areas of trees. Whip and tongue grafting are being used for reproducing the nursery stock.

Budding is a method of propagation that is similar to grafting. It is popular and even successful method that is used to reproduce fruits and roses and sometime nut trees. Budding basically consists of a removing bud from one plant and then placing them on the stem that is found to be closely related to plants.

Normal buds are those bud that are developed normally on the plants at the terminally or axis on the stem. Normal buds and the adventitious buds (the buds that are developed in the position other than the normal buds that are found) are used for budding.

Plant breeders are using budding as a means of changing cultivars of the plant species. Some of the common budding methods like patch budding, t-budding and chip budding. Tissue culture or micropropagation is considered as a special type of asexual reproduction.

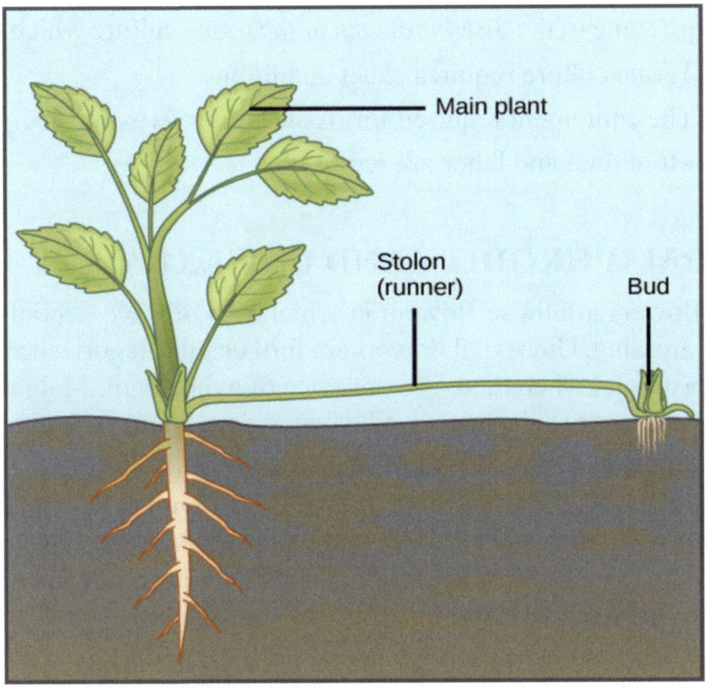

Figure 2.5. An image representing the method of budding.

Source: Image by Wikimedia Commons

Tissue culture involves placing a very small piece of plant tissue on a specialized culture medium used in a test tube. The plant tissue is found to multiply and grows into a new plant that is under sterile conditions in the test tube.

Plant tissue is being used in the tissue culture which includes the tips of shoot and root, and some of the sections of leaves and stems. Other tissues often include the embryos, pollen grains and the seeds. There are some advantages that exist for using the micro propagation:

- Tissue culture is a cost-efficient method of reproducing plants.
- Tissue culture allows a large number of offspring that can be produced quickly.
- Plants reproduced by tissue culture have the same characteristics as the parent plant.
- Tissue culture allows growers to produce plants that are free of disease.

There are some of the disadvantages of the tissue culture which includes:

- Tissue culture requires exact conditions.
- The equipment required for tissue culture is expensive.
- More time and labor are required.

2.5. HERMAPHRODITY AND UNISEXUALITY

Unsexual flowers are those flowers in which one of the essential parts are found to be missing. Unisexual flowers are further subcategorized as a female flower or pistillate, when there is a presence of gynoecium. Male flowers or staminate flowers are subcategorized when androecium is present.

It is believed that different families are having different types of flowers. Legumes are considered to have a bisexual flower with the petals which are modified into the banner petal, keel, and wing petals. Cucurbits are usually considered to have unisexual flowers but sometimes can have bisexual flowers too, like in case of watermelons.

There is a specific terminology that is used for the plants that are based on the types of flowers they bear and which type of flowers are existing on each of the plant. The plant is called as a bisexual or hermaphrodite if it is bearing only the bisexual or hermaphrodite flowers.

The plants that are bearing the unisexual flowers are further being subcategorized as monoecious if both the male and female flowers are occurring on the same plant and dioecious only if male and female flowers occur on different plants.

There is another condition that is called sub-dioecy that may also occur. Under this category, the plants are being subcategorized as an andromonoecious if both the male and hermaphrodite flowers are present. Gynomonoecious is if both the female and the hermaphrodite flowers are found. Whereas, the trimonoecious is when female, male and hermaphrodite flowers are borne on the same plants.

2.6. SELF-POLLINATION AND CROSS-POLLINATION

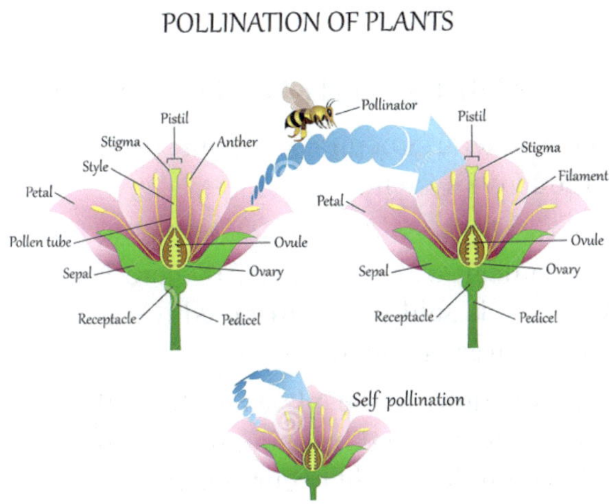

Figure 2.6: Pollination methods.

Source: Image by Dreamstime.com

Plants reproducing sexually can be subcategorized further which is based on the source of the pollen that is pollinating the plant. Self-pollination basically occurring when the pollen from the flower pollinates the stigma of the same flower or can be of another flower present on the same plant.

Cross pollinated species are those species where the pollen from the flowers gets transfer the stigma of a flower from one plant pollinates to another plant. Stebbins, 1950 often observed that there is a relationship between the length of the lifecycle of a plant and for its reproduction mode.

Annual plants are considered to have fewer opportunities for the genetic recombination which is in their short life span, self-pollination is the important key to reproductive assurance (Duvick, 1966).

On the other side, the perennials are considered to be most trendy towards the outcross because it is found that they have more opportunities towards genetic recombination in a life span spread over many years (Duvick, 1966).

The term self-pollination and cross -pollination: are considered as one of the methods of pollination which is more predominant than the other in the crop because there is some amount of outcrossing in the self-pollinating crops and in the Selfing in the cross-pollinating crops which commonly occurs.

2.7. HOMOZYGOSITY AND HETEROZYGOSITY

It is believed that the genetic structure of a plant species is being determined largely by its reproductive system.

In the case of species reproducing asexually, the offspring are found to be genetically identical to their parents. If there is any variation in the asexual progeny then it is said to be attributed towards the environmental effects or towards the rare genetic mutation.

Vegetative reproduction of the plants is basically heterozygous and their heterozygosity is being fixed by using the clonal propagation because there is no recombination that might occur and all the progeny is essentially arising from the same plant.

In the case of species using sexual reproduction, there are two kinds of mating that are possible: self-pollination and cross -pollination. As there is no possibility for gene recombination in self-pollination, there might be an exception to the occasional event of outcrossing.

In the case of species that are self-pollinating, there is an observation of variation which is more common among the populations than within the populations. This trend is being reported in the case of *Leavenworthia* of the Brassicaceae family (Charlesworth, 1998; Liu et al., 1999).

Figure 2.7: Leavenworthia.

Source: Image by Flickr

It is being observed that the variation in the self-pollinating species is comparatively greater than in the case of cross-pollinating species. The genetic structure of a species is said to give further influences on the adaptability of that species.

It is believed that the wider genetic base of the cross-pollination species is giving them better- buffering capacity that helps them to survive various abiotic and biotic stresses when it is compared to the self-pollinating species. The idea is being supported by the experiment conducted by Steven, 1948; at the time the Steven estimated that the crop losses are due to the disease occurring in the different reproduction systems.

The result often suggested that there is maximum disease loss that is occurring in the asexual propagating species which is being followed by the self-pollinating, and finally by the cross-pollinating species.

Therefore, in the case of cross-pollinating crops, there is continuous self-pollination that has an adverse effect in the form of inbreeding depression. This basically occurs due to the accumulation of the deleterious recessive alleles. These are expressing in the homozygous state which is basically in the selfed plants of a cross- pollinated species.

It is observed that the self-pollinated plants are not facing this problem because of the continuous selfing that takes place over many generations and are also found to be deleterious recessive alleles that get purged. It is believed that the mode of reproduction is also influencing the genetic structure of the population.

Self-pollination is often increasing the homozygosity due to which the accumulation of similar alleles is resulting from the selfing which is over several generations. Whereas, the process of cross- pollination is increasing the heterozygosity which is basically due to the frequent recombination and segregation.

So, the genetic structure of a self-pollinated population is said to be heterogeneous with the homozygous individuals, and of a cross- pollinated population it is observed to be homogenous with the heterozygous individuals.

2.7.1. Inbreeding

It is believed that the self-pollinating crops and the cross-pollinating crops are responding differently towards inbreeding. Generally, the inbreeding method is the natural mode of breeding used in the self-pollinated crops and it is believed to produce desirable results by simply increasing the homozygosity and uniformity of the plants.

There is no adverse effect of inbreeding on the self-pollinated plants because of the continuous selfing that takes place over the generations, and the population which has been purged of the recessive deleterious alleles.

Therefore, there is the enforcement of inbreeding in the naturally cross-pollinating species, as they may lead towards the drastic consequences. The simple illustration of the adverse effect of inbreeding can be seen simply by the results of the independent inbreeding experiments that take place in corn which was conducted by East, 1908 and Shull, 1909.

Allard, in the year 1960 often summarized the most important effect that was because of the inbreeding which was reported by these investigators are as follows: it basically starts with the appearance of the several lethal and the type of sub-vital genes which is in the early generations of the selfing. This was followed by the separation of the population into the distinct lines, which is often said to become increasingly uniform within the and which is increasingly distinct from the other lines over the generation of a selfing.

It has been observed that many of these lines are showing general decrease in their fertile and vigor and is found to be difficult to maintain it even in the most favorable cultural condition.

2.7.2. Hybrid Vigor/Heterosis

Hybrid vigor is a phenomenon of increased vigor in the hybrid when compared to both of its parents. This phenomenon has come into knowledge in the 20th century in corn crop F1 hybrids. It is believed that the resulting plant had a higher growth rate, which was phenotypically considered as superior and had also increased in the yield in comparison to their parents.

As it is clear, that the whole country cannot adopt one hybrid, therefore there is a presence of different hybrids that had been developed in order to adapt to different areas. The basic mechanism that is being proposed in order to involve in heterosis is considered to be dominance and overdominance.

Dominance basically means that the dominant allele masks are the effect of the recessive allele. Whereas, the overdominance means that the combination of the genotypes from the two different parents which is leading towards an increase in the vigor (Chahal and Gosal, 2002).

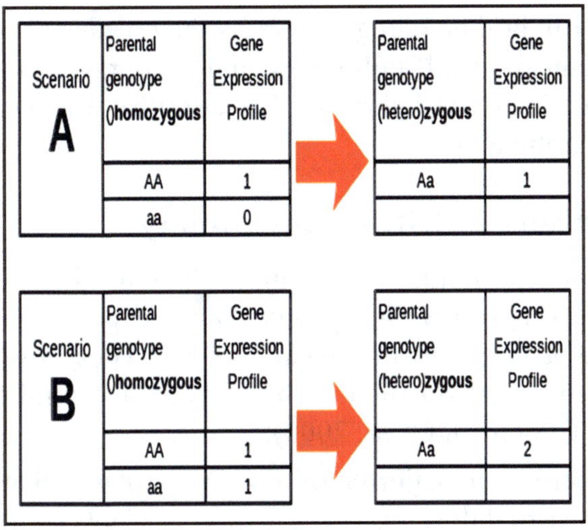

Figure 2.8: Heterosis tables.

Source: Image by Wikimedia Commons

2.8. SELF-INCOMPATIBILITY

In flowering plants, Self-incompatibility (SI) has brought an evolutionary advantage. The advantage of self-incompatibility (SI) is the effectiveness of avoiding the inbreeding and encouraging outcrossing. With the help of Self-incompatibility, recent years have seen the creation of many vegetable and fruit hybrid cultivars.

Another advantage is its ability to produce F1 hybrids using two SI lines. It works as a parental component and helps in the elimination of the laborious emasculation of the female parent. One of the most important objectives of the breeders in modern plant breeding is the F1 hybrids. In the USA, 1921, the first F1 hybrid breeding system was established. The establishment was done in corn and came to be known as 'hybrid corn'.

Because of the laborious emasculation of the male flower (tassel), the problem emerged and lead to the high cost of the seeds (Franklin-Tong, 2008). In order to prevent themselves from self-pollinating, the mechanism of self-incompatibility (SI) is utilized by flowering plants (angiosperms) (Silva and Goring, 2001).

This resulted in the utilization of self-incompatibility for achieving a more efficient F1 hybrid breeding system by the breeders. Since SI prevents the flowering plant from self-pollination and encourages outcrossing, there can be the production of F1 hybrid seeds readily from two parental SI lines with the help of pollinators.

2.8.1. Mechanisms of Self-Incompatibility

There are numerous unique and distinguished mechanisms for SI and being evolved through flowering plants. The flowering plants have some of the species that produces unisexual flowers. Those flowers are either male/staminate or female/pistillate. This acts as a natural barrier to self-fertilization (McCubbin and Kao, 2000).

However, there is a majority of flowering plants that have perfect/hermaphrodite/bisexual flowers. These flowering plants contain both female reproductive organs and male reproductive organs in close proximity on the same flower (Kao and McCubbin, 1996).

There is a broad categorization of SI into heteromorphic SI and homomorphic SI. Heteromorphic self-incompatibility is basically the SI of flowering plants due to some morphological barriers in flowers. This occurs in some hermaphroditic flowering plants.

Those produce the plants that have structurally distinct reproductive organs. For instance: thrum flowers with long stamens and a short style or pin flowers with short stamens and a long style (e.g., Primula, Oxalis).

There is a problem of a topological barrier to self-fertilization due to the relative positions of reproductive organs (Ebert et al., 1989). The homomorphic SI is another form of SI in which the genetic mechanism is responsible for the avoidance of self-fertilization.

On the basis of the type of genetic mechanisms involved. There is another subcategorization of SI into gametophytic SI (GSI) and sporophytic SI (SSI).

Gametophytically controlled SI is considered as the most widespread SI system. Usually, the SI is controlled by a single S locus containing a multiple S-alleles (Franklin-Tong and Franklin, 2003). There are two types of loci found in some species such as in grass (Franklin-Tong and Franklin, 2003). In the GSI system, the pollen and its SI phenotype are determined by its own haploid (gametophytic) genome.

However, in the SSI system, the determination of the pollen SI phenotype is done by the diploid genome of its parent (sporophyte) (Silva and Goring,

2001).

2.8.2. Self-Incompatibility as an Important Trait in Plant Breeding

In plant breeding, one of the advantages of the SI is its promotion of outcrossing. The F1-hybrids have great economic importance in the number of modern plants. The importance of the SI being their uniformity and hybrid vigor (heterosis) in a number of crops (Franklin-Tong, 2008).

On a large-scale production of F1-hybrid seed is the efficient method of its production by utilizing SI. Also, in order to avoid contamination by self-pollinated seeds, it is essential to have a SI system (Franklin-Tong, 2008). However, the main problem in the utilization of SI is the establishment of two pure line parents in SI plants.

For finding the solution to this problem, the breeders chose to have SI lines from a large number of seed stocks. Further, utilizing it for the physiological or genetic breakdown of SI for the production of pure lines as SI is not always stable (Franklin-Tong, 2008).

However, the problem of SI could be solved through the various external and physiological conditions such as treatment with CO2 gas, irradiation, and pistil grafting (Nettancourt, 2001).

2.9. CONCLUSION

Reproduction in plants is a process of production of new offspring. Therefore, all the organisms multiply and are reproducing their own kind. In plants, there are basically three modes of reproduction: vegetative, asexual, and sexually.

Vegetative reproduction is a process that takes place through bud formation. Which means that the new develop from their own parents in the form of bud, or organ. The asexual process is a similar method which produces the new one which is similar to their parents. That does not form by two different parents.

Whereas, sexual reproduction is a process which involves two different gametes that is, male and female gametes combine to form an individual. There are different asexual reproduction methods used for reproduction. These types are: fragmentation, spore formation, budding and vegetative propagation.

Flowers are considered as the reproductive parts of a plant. Flowers are usually found to be unisexual with either the male or female reproductive parts. A bisexual flower is considered to have both the male and the female reproductive parts.

The male gametes are found in the pollen grains and female gametes are found in the ovule. Pollination is a process in which the transfer of pollen grains takes from the anther of one flower get to the stigma of the same flower or another flower.

The pollination process is basically of two types, first is self-pollination and second is cross- pollination. In self-pollination, the pollen grains are being transferred from the anther to the stigma of the flower. In the process of cross- pollination, the pollen grains are being transferred from the anther of one flower to the stigma of another flower of the same kind.

Pollination process takes place in the plants through the water, wind, and insects. The fusion of the male and female gametes is known as the fertilization. The fertilized egg is known as a zygote. Zygote further develops into an embryo.

Fruits are considered as the mature ovary, whereas the ovule develops into a seed, which is said to contain a developing embryo.

Therefore, the reproduction process takes place in all the plants, especially through sexual reproduction, but this process also takes place through vegetatively and asexually, which depends on the variety of plant.

REFERENCE

1. 2020. *Reproduction In Plants*. [ebook] Available at: <http://ncert.nic.in/textbook/pdf/gesc112.pdf> [Accessed 26 May 2020].
2. Boavida, L., Vieira, A., Becker, J. and Feijo, J., 2005. Gametophyte interaction and sexual reproduction: how plants make a zygote. *The International Journal of Developmental Biology*, [online] 49(5-6), pp.615-632. Available at: <https://www.researchgate.net/publication/7663452_Gametophyte_interaction_and_sexual_reproduction_How_plants_make_a_zygote> [Accessed 26 May 2020].
3. Gill, R., MacAllister, C. and Zhang, T., 2013. *4. Plant Reproductive Systems - Plantbreeding*. [online] Plantbreeding.coe.uga.edu. Available at: <http://plantbreeding.coe.uga.edu/index.php?title=4._Plant_Reproductive_Systems> [Accessed 26 May 2020].
4. Lambers, H., n.d. *Plant Reproductive System - Physiology Of Plant Reproduction*. [online] Encyclopedia Britannica. Available at: <https://www.britannica.com/science/plant-reproductive-system/Physiology-of-plant-reproduction> [Accessed 26 May 2020].
5. Muthoni, J., Shimelis, H. and Melis, R., 2014. Genetics and Reproductive Biology of Cultivated Potato (Solanum tuberosum L.). *Reproductive Biology of Plants*, [online] pp.164-194. Available at: <https://www.researchgate.net/publication/260174574_Reproductive_Biology_of_Plants> [Accessed 26 May 2020].
6. n.d. *Life Cycle Patterns*. [ebook] Available at: <https://www.cabrillo.edu/~ncrane/bio1c/botPDFs/Labs/bacteria,%20cells,%20protists,%20lifecycles/Life%20Cycle%20Patterns.pdf> [Accessed 26 May 2020].
7. n.d. *REPRODUCTION IN PLANTS*. [ebook] Available at: <http://visionpointnios.co.in/courses/314/Lesson-19.pdf> [Accessed 26 May 2020].
8. n.d. *REPRODUCTIVE MORPHOLOGY OF FLOWERING PLANTS*. [ebook] Available at: <http://courses.washington.edu/bot113/summer/LabExercises/Repromorphhigh.pdf> [Accessed 26 May 2020].
9. n.d. *Sexual And Asexual Reproduction Of Plants*. [ebook] ims. Available at: <http://gibsonburgagdept.weebly.com/uploads/2/3/5/5/23557960/sexual___asexual_repro_of_plants_-_pkt.pdf> [Accessed 26 May

2020].

10. Schultz, S., 2020. *Reproduction In Plants - Biology Encyclopedia - Cells, Body, Examples, Function, Process, Animal, Different, Organisms, Chromosomes, Organs*. [online] Biologyreference.com. Available at: <http://www.biologyreference.com/Re-Se/Reproduction-in-Plants.html> [Accessed 26 May 2020].
11. Allard, 1960
12. Chahal and Gosal, 2002
13. Charlesworth, 1998
14. Duvick, 1966
15. East, 1908
16. Ebert et al., 1989
17. Franklin-Tong and Franklin, 2003
18. Franklin-Tong, 2008
19. Kao and McCubbin, 1996
20. Liu et al., 1999
21. McCubbin and Kao, 2000
22. Nettancourt, 2001
23. Shull, 1909
24. Silva and Goring, 2001
25. Stebbins, 1950
26. Steven, 1948

Chapter 3

Fertilization and Double-Fertilization

CONTENTS

This chapter is about the process that is used in the flowering plant for reproduction. The process of fertilization and double fertilization is being explained in this chapter. Basically, the different type of fertilization that is used by the flowering plants.

There are basically three type of fertilization process known as Porogamy, Chalazogamy, and the Mesogamy that are process through the plants proceed their fertilization process. There are certain steps that takes place during the process of fertilization. Some of the steps are gamete attachment, sperm recognition, gamete activation, gamete fusion and karyogamy.

These steps are taking place whenever there is process of fertilization in the flowering plant. This chapter also includes the process of Double fertilization. Double fertilization is the process that takes place when the female gamete fuses with both the sperm and forms polyploid. Similarly, the discovery or the origin and few other points about the double fertilization is explained below.

3.1. INTRODUCTION

Generally, fertilization is the fusion of haploid gametes that are used for initiating the development of a new diploid organism. Normally in the case of animals, an ovum or egg fuses with a motile sperm in order to generate the diploid zygote, which will further develop into an embryo.

According to the current representation of early branching of the eukaryotes which includes fungi and protozoa, are considered to produce the mating types of the opposite sex that are generating a zygote after fusion.

Likewise, the gametes of the flowering plants are not the direct product of the meiosis. There is a formation of male gametes that are formed after two additional mitotic cell division, which in return creates three-celled male gametophytes, the pollen and pollen tube.

The female gametophytes or embryo sac that is developed within the maternal organ which is called the ovule, developed after the three additional mitotic nuclear divisions that take place on functional megaspore. Cellularization is a seven called embryo sac which is being established in the majority of the angiosperms.

Angiosperms have two female gametes known as eggs and central cells, and accessory cells known as synergids and antipodal, which are present at both the poles of the embryo sac (Drews et al. 2011; Spunk et al.,2011).

In order to reach the two female reproductive cells, the sperm cells present in the angiosperms have to conquer distances of about a few millimeters in the species such as Arabidopsis or up to 30 centimeters in case of maize. While the mosses and ferns are still possessing the motile sperm, the sperm cells of the angiosperm might have lost their motility.

Whereas, the pollen tube cells are acting as the vehicles that are used to transport the sperm pairs deep through the maternal reproductive tissues. The tube cells basically grow at its tip with a speed of approximately 1 cm per hour with the cargo at a short distance from the tip.

It is observed that during their journey, the sperm cell is found to be connected to each other and to the nucleus of the tube cell, which is moving as a 'male germ unit'.

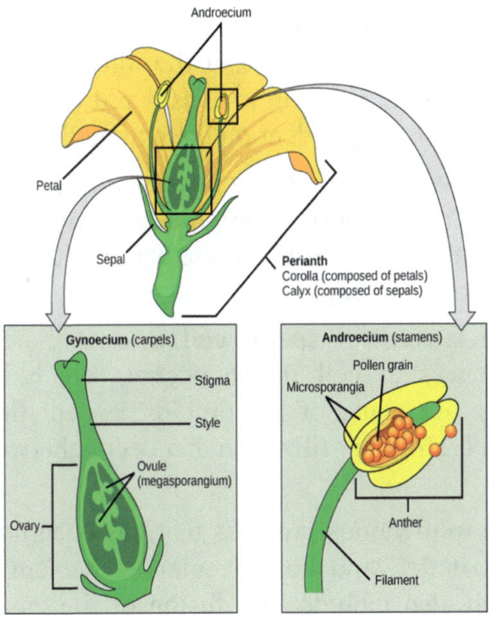

Figure 3.1: An image representing male reproductive and female reproductive part of a plant.

Source: Image by Wikimedia Commons

It is believed that intensive communication often takes place during the journey of arduous pollen tube which occurs within the maternal tissues of the stigma, transmission tract, ovule, and style. These processes are collectively referred to as a program phase that has been reviewed recently.

Once the sperm arrive at the ovule, the directive communication is said to continues with the female gametophyte (especially the synergid cells), that guides the pollen tube by using the micropyle opening of the ovule and even help them by regulating the release of their contents (Higashiyama et al., 2015; Bleckmann et al., 2014; Dresselhaus et al., 2009).

It is being observed that micropyle opening of the flowering plant ovules enables the sperm to access and are reminiscent of the micropyle that is found in many of the insects and fish eggshells.

In the funnel-shaped micropyle of the herring and flounder eggs, asperm gets attracted around the opening. Them micropyle detects the motile sperm into the micropyle and across the chorion it gets attached to the oocyte's plasma membrane.

There is a restriction in the number of sperm entering, this is being done by the diameter of the inner aperture of the micropyle. The attraction of the sperm to the micropyle opening is considered to appear to be species-specific and is found to be dependent on the extracellular Ca^{2+}.

Therefore, it shows the strong functional overlap with the pollen tube that guided it in the plants. During the process of pollen tube reception in the flowering plants, the two sperm cells are being released towards the cleft between the central cell and egg.

Only after the successful sperm cell adhesion, activation and cell positioning the process of cell fusion occurs, which is also known as plasmogamy. There are some of the orchids, except the Podostemaceae family, where there is single fertilization that occurs between the sperm and egg.

There has been tremendous progress that has been made from the past five years in order to understand the molecular mechanism of the fertilization in the plants, which also includes the fusion of the gamete nuclei called karyogamy.

With that, it is also preventing the polyspermy which is a fusion of multiple sperm cells with the female gametes, by simply avoiding the attraction of the multiple pollen tubes known as polytubey.

According to recent research which was using in vitro fertilization , there are early steps of fertilization that are sharing some homology especially with few animal species.

The process of double fertilization of the flowering plants was discovered many years ago. The cytology of the gametes is now well known. However, the description of the fertilization steps are still poor and even many of the cellular and molecular mechanism are there which are unknown.

Therefore, the process of gamete formation is being followed by the cytosolic calcium increase which takes place in the fertilized egg as well as a calcium influx. Further, the understanding of the fertilization is said to come from the analysis of the mutants that are being isolated from *Arabidopsis thaliana*.

3.2. FERTILIZATION

Fertilization is a process of sexual reproduction that takes place in the plants. This occurs after the process of pollination and germination. Fertilization can be also defined as the fusion of male gametes known as pollen, with the female gametes known as an ovum, in order to form a diploid zygote.

Figure 3.2: Fertilization process by the bees in the flowering plant.

Source: Image by Wallpaperflare

Fertilization is a physiochemical process that basically occurs after the process of pollination of the carpel. This complete process of fertilization takes place in the zygotes in order to develop into a seed. It is considered that in the fertilization process, the flowers play a significant role, as they are considered as a reproductive structure of the angiosperms (flowering plant).

The process of fertilization in the plants basically occurs when the gametes in the haploid condition get fused to form the diploid zygote. In the process of fertilization, it is believed that the male gametes get transferred into the female reproductive organs with the help of pollinators.

These pollinators can be birds, honey bees, butterflies, bats and flower beetles. This will result in the final product, that is, the formation of the embryo in the seed.

3.3. TYPES OF FERTILIZATION

Fertilization is a process that can be grouped into the following three types. Fertilization process is classified on the basis of the entry of the pollen tube into the ovule.

3.3.1. Porogamy

Porogamy is considered as the common type of fertilization that is basically carried out in the angiosperms or in the flowering plants. In this type of fertilization, the pollen tube is said to enter the ovule through the help of micropyle.

3.3.2. Chalazogamy

Chalazogamy is the type of fertilization which is being carried out in all the species of casuarina. In this type of fertilization process, the pollen tube enters the ovule by using the pollen tube.

3.3.3. Mesogamy

Mesogamy is the type of fertilization that is basically seen in all the Cucurbits. For example, ridge gourd, bitter gourd, pumpkin and other gourds. In this type of fertilization, the pollen tubes basically enter the ovule through the middle part or through the integuments of the ovule.

3.4. FERTILIZATION PROCESS

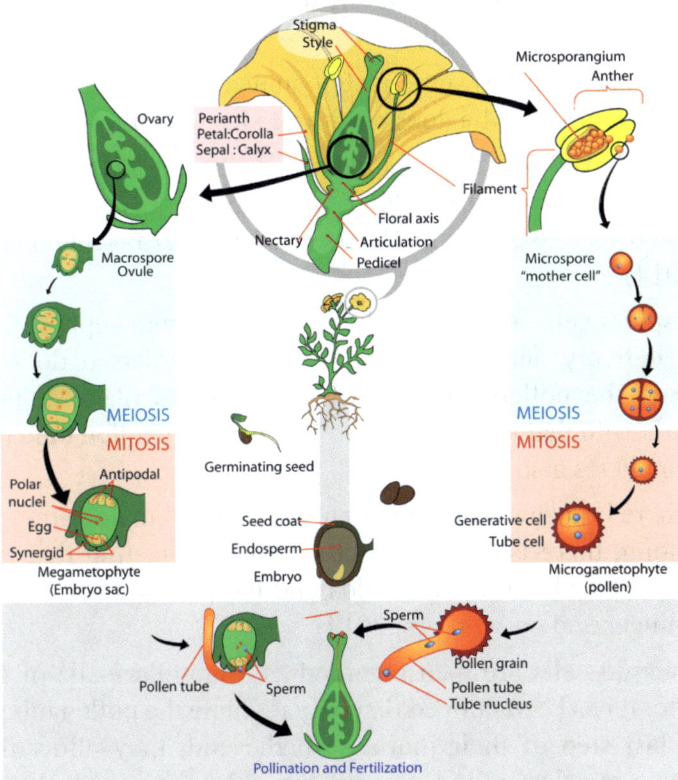

Figure 3.3: An image representing the process of fertilization.

Source: Image by Simple Wikipedia

3.4.1. Pollen Tube Reception and Sperm Cell Delivery

After the arrival of the sperm cells at the female gametophyte and after the interaction with the receptive synergids cells, the pollen tube is said to burst and release its contents. This content is basically two sperm cells, while the receptive synergid cells are considered to degenerate.

Therefore, this highly coordinated lysis of the two gametophytic cells is considered to expose the sperm cell cargo that is released to the eggs and central cell. Further, it has been observed that the pollen tube enters the receptive synergids cell with the help of the filiform apparatus.

The filiform apparatus is a unique structure of membrane invaginations and then there is a formation of a highly thickened cell wall at the micropylar region of both the synergid cells.

However, according to some recent studies, it becomes evident that in Arabidopsis, the pollen tube does not invade into the filiform apparatus.

But it is believed that it enters the embryo sac at the micro distant site, simply by growing around or along with the receptive synergid cell but it gets bursted to cease growth (Leshem et al., 2013; Denninger and co-workers, 2014).

As the sperm cells are not motile, it is necessary to support the targeted sperm cell delivery and gamete attachment. In order to do so, there is redirecting of the pollen tube growth which takes place at the filiform apparatus and with that the proximity of the bursting pollen tube tip towards the female gametes also takes place.

When there is an entry of the pollen tube in the female gametophyte and before bursting, there is intensive extracellular signaling that takes place between the one or both the synergids and the pollen tube (Leshem et al., 2013; Denninger and co-workers, 2014).

The synergid cells are then regarded as a granular cells of the female gametophyte. It has been noticed that they regulate the pollen tube attraction during the last step of their journey. In the end, they allow the growth arrest and sperm release, and prevent the polytubey upon the successful fertilization.

In Arabidopsis, the search conducted for the mutants defective in pollen tube growth arrest and the burst resulted in the identification of the FERONIA/ SIRE′NE (FER/SRN) encoding a member of the Catharanthus roseus which is a subfamily of a receptor-like kinase CrRLK1L-1.

FER can be broadly expressed and with that the encoded protein that is localized at the plasma membrane present in the diverse cell types. It is believed that in the Arabidopsis ovule, FER-GFP is appearing most prominently at the filiform apparatus.

Its extracellular domain is considered to be likely glycosylated as a mutant defective in the genes EVAN (EVN) and TURAN (TUN) which are encoding a uridine diphosphate (UDP) glycosyltransferase superfamily protein and a dolichol kinase.

EVN and TUN both are required for the N-glycosylation process that takes place in the endoplasmic reticulum (ER) and are displaying the same for the mutant phenotype.

Pollen tube reception and synergids degeneration are also found to be affected in the Arabidopsis mutant LORELEI (LRE). LRE is found to be encoding a GPI-anchored protein which is expressed predominantly in the synergids cells.

According to recent work that was done on LRE and on LORELEI-like GPI-anchored protein 1 that is closely related is being suggested that both the proteins are found to critical for the cell surface signaling capacity of the FER.

LRE and LLGI are found to interact with the FER and is also observed to play an important role in facilitating the localization of the FER towards the cell membrane or even stabilizing the FER-receptor complex at the cell surface.

While comparing the extracellular domain of the FER with the corresponding proteins that are present on the other species of the Brassicaceae family are further considered to reveal the high degree of the amino acid changes in the domain as compared with the kinase domain.

This is further found to indicate the domain that may be under the selective pressure for the species-specific interaction that takes place with the pollen tube and therefore contributes to the reproductive isolation between the species.

The complementary pollen tube-derived ligand is considered that they may also interact with the FER that has not been identified. The small secreted Cysteine-Rich Peptide (CRP) of the RAPID ALKANIZATION FACTOR (RALF) which is a sub-family of CRPsis believed to be shown directly by interacting with the FER. Thus, interaction is resulting in the process of phosphorylation of the plasma membrane- localized H+ -ATPase 2 that mediates the inhibition of proton transport.

3.4.2. Gamete Recognition and Attachment

When compared with the animals, there are very few studies known of the direct gamete interaction that occurs in the flowering plants.

During the process of pollen tube journey, there has been a number of mechanisms that are being established in the plants in order to distinguish between the self or non- self and alien pollen for preventing the inbreeding and fertilization with the foreign sperm cells.

Sperm cells are being delivered towards the site of the fusion of the gametes that takes place by the pollen tube. Therefore, one would expect that the flowering plants might have lost their specific gamete recognition systems.

Rapidly evolving cell surface proteins are considered to play a key role during the process of gamete recognition and adhesion in the metazoans and they even reinforce the barriers towards the interspecific hybridization.

The fact is that the plants have many reproductive hurdles that are before the gamete interaction and it should not be interpreted to the means that the plants do not have species -gamete recognition mechanisms.

There is a unique mechanism that is used in the plant gamete recognition which basically remains and that is needed to be understood about how one sperm attaches to and gets fuses with the egg cell. While the second sperm cell which almost simultaneously fuses with the central cell.

With the exception of a few plant species, are considered to contain a dimorphic sperm cell. In these dimorphic sperm cells, there is a larger sperm cell with more mitochondria and fuses with the central cell and the smaller sperm cell with the egg cell.

Sperm cells in most of the flowering plants are considered to be isomorphic and are found to be preferential fusion with the female gametes that has not been observed. In Arabidopsis, it has been shown that the sperm cells are able to fuse with the female gametes. Thus, it is considered to be preferential fertilization does not exist in these species.

Live cell imaging during the process of double fertilization is being further revealed that the two sperms cells basically remain close together and are relatively immobile after being delivered to their future fusion sites.

The beginning of the sperm nuclei movement is said to occur around 7.4 minutes after the burst of the pollen tube often suggested that the plasmogamy has taken the place. However, it is believed that it remains to be explored to the extent that gamete recognition and adhesion often take place during a short period of time, even though the male and the female gametes communicate with each other.

According to a study, the Arabidopsis sperm cells often become competent for fertilization after they are released from the pollen tube. This often indicates that the communication takes place between the gametes which are following the sperm cell that is being released from the pollen tube, but this process takes place before the execution of the fusion process.

The composition of the protein that takes place on the surface of the plant gametes has not been determined. But, the experiment of lectin labeling by using the isolated male and female maize gametes often indicates the presence of the surface carbohydrates. This might play an important role for cell adhesion.

The first indication that can be used for the surface proteins that are involved in the gamete attachment and recognition existing in the flowering plants, which is similar to metazoan and algae, is often given by the functional characterization of the Arabidopsis Gamete Expressed 2 (GEX2). GEX2 is a protein that is localized to the sperm cell surface.

GEX2 also contains a single pass Transmembrane Domain (TMD) which is present at the carboxyl-terminal region and is also considered as an extracellular localized filamin repeat domain which is similar to the immunoglobin-like (Ig-like) domain. This domain is seemed to be involved in the gamete interaction that takes place in the mammals and in the non-seed plants like algae (Evans, J.P., 2012).

By using an *in vivo* assay, it basically shows that GEX2 often requires for both the sperm-egg cell and sperm-central cell attachment.

By investigating the GEX2 of the closely related species, *Arabidopsis thaliana* and *Arabidopsis lyrata* often revealed that the filamin repeat domain is found to be evolved much faster than other regions of the proteins. This makes it an attractive domain that can be used to mediate the species-specific gamete attachment.

3.4.3. Gamete Activation Before Fusion

It is considered that there is a couple of minutes that is often required in Arabidopsis after the pollen tube before bursting of the sperm cells to fuse with the female gametes (Hamamura et al., 2011). Until recent times, it was basically unclear about what exactly is happening to the gametes during this time.

According to the recent breakdown, it was noticed that of a family of small cysteine-rich proteins which was named EC1 (ECG CELL 1). EC1 often accumulates in the storage of the vesicles in the egg cell before the process of fertilization.

EC1-GFP fusion proteins are being detected extracellularly only after sperm delivery. This has been suggesting that exocytosis of the EC1-GFP is being actively triggered. Both the gamete fusion is being impaired in the EC1 deficient ovule and EC1 peptides, which is exogenously applied to the sperm cells.

It stimulates the relocation of the potential fusogen Generative Cell-Specific 1/ Hapless 2 (GCS1/HAP2) from the endomembrane system to the plasma membrane. This indicates that the egg cells secrete the EC1 proteins in order to achieve rapid sperm activation and gamete fusion.

It has been shown later, that there are some of the sperm cells present in the EC1 deficient ovules which are able to accomplish markedly the delayed fusion that occurs with the female gametes. It has been suggesting that these sperm cells are having a very slowly acquired fusion competence.

EC1 related proteins are considered to be expressed in the synergids cells, which may also contribute towards the gamete activation during the process of the sperm cell delivery.

Altogether these findings are considered to indicate that the first egg cell becomes activated during or immediately after the sperm cell release, in turn, activating the sperm cells and with that it also enables them to fuse with them quickly.

However, the plasmogamy of the sperm cell is said to devoid of the Cyclin Dependent Kinase A1 (CDKA;1) with the central cell which is found to be sufficient to trigger the mitotic division despite of having the failure of the karyogamy (Which is a fusion of the gametic nuclei). This was also suggesting that the signaling even which is associated with the sperm entry into the central cell like in most of the animal fertilization events.

Notably, both the plant female gametes are displaying transient Ca^{2+} during the process of sperm cell discharge from the pollen tube which is prior to the fusion of the sperm-female gametes. This was offering the possibility about the first level of the female gamete activation which is being achieved by the contents of the pollen tube and not directly by the sperm.

According to the recent work that was done in the fruit fly named as *Drosophila melanogaster*, it has been demonstrated that the Ca^{2+} rise that basically occurs in its oocytes which are considered to occur independently of the sperm fusion. Instead of this, they rely on the mechanical pressure on the egg which is provided normally during the egg laying process or by experimentally by an investigator.

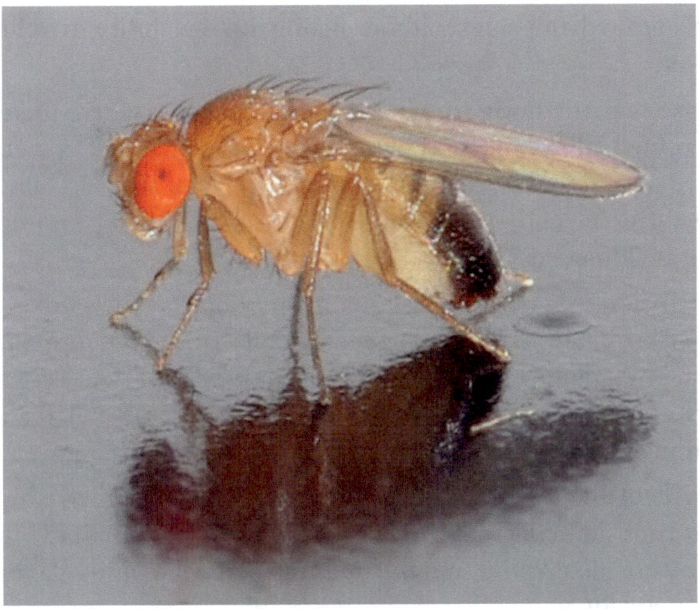

Figure 3.4: *Drosophila melanogaster.*

Source: Image by Flickr

In *Drosophila*, the sperm-egg fusion process occurs after the egg transit take place through the oviducts. But it can be said that it is the only stimulus for the Ca^{2+} wave that occurs in all the animal egg tested. The passage through which the female tract at egg -laying gets finalized by the activation of the event by the mechanically induced wave of Ca^{2+}.

Similarly, the bursting of the pollen tube often causes the deformation of the female plant gametes. Thereby, they might stimulate the mechanosensitive Ca^{2+} channels similar to those seen in Drosophila.

This hypothesis is being supported by the observation that basically tells about both the female gametes which respond towards the pollen tube/ synergid cell burst with a single Ca^{2+}transient, which is being observed in a gamete fusion mutant.

3.4.4. Gamete Fusion (Plasmogamy)

There are very few cells that are present in the plant or animal which are normally programmed to fuse with another. In fact, it is observed that the animal cells pride themselves without fusing, on the ability to tightly adhere to each other.

This case is particularly true in the epithelial cells, where there are tight junctions, robust adhesions, and close applications that take place between the cells are considered as the norm but cells do not fuse. In humans, with about 200 different cell types where only a few cell types are normally undergoing fusion events.

These fusion events are: skeleton myoblasts (myogenesis), eggs and sperm (fertilization), some placental cells (placenta formation in the mammalian implantation), osteoclasts (bone remodeling). As a consequence, these sparse examples are considered as knowledge of cell fusion events which is limited.

It is evident that the specific membrane proteins which are called as fusogens are considered to require in order to modulate the lipid bilayers and thereby, they are also facilitating the merging since the lipid bilayer does not fuse spontaneously.

The roundworm known as *Caenorhabditis elegans* is considered to undergo significantly with the cell fusion in its outer, dorsal epithelium. This roundworm often requires the type I membrane protein EFF-1 to complete the process.

EFF-1 driven fusion is considered to entail the trans-trimerization in order to bring the TMDs which are anchored in the two opposite membranes in the close contacts. It was being reported that the adding of EFF-1 and other fusogens to the cells can stimulate cell fusion in the cells that basically do not normally undergo the process of fusion.

Therefore, there was a major argument that was related to the cells that do not fuse because they do not possess a fusogen and not because they are molecularly recalcitrant to fusing. It is considered that a plethora of the membranous viruses is often able to fuse with the diverse animal cells and also supporting this contention.

It was surprising that there are few and diverse proteins that are known to play an essential role in the gamete fusion, despite they are being such a fundamental and ancient process. There are several excellent review articles that are on the fusion-essential proteins that are being discovered so far in the animal's species.

However, in the mammals, the only essential factors that are required for the fusion is being described to the date are sperm IZUMO, and the ovum Juno and CD9. IZUMO1 is considered as a type 1 membrane glycoprotein with an extracellular Ig-like domain, one of the Izumo domain is at its amino terminus and a short cytoplasmic tail.

IZUMO1 is on the inner and the outer acrosomal membrane in the mature spermatozoa and even translocating it to the equatorial segment only during the acrosomal reaction. The equatorial segment is the region of the acrosome reacted sperm that is considered as important in initiating the fusion with the egg plasma membrane.

Recently, the eggs are being expressed as a receptor for the IZUMO which was identified as the GPI anchored folate-receptor protein Juno, which is also known as Flor4. Upon the IZUMO1-Juno interaction, the fusion is essential tetraspanin CD9, which is being recruited to the adhesion site, often suggesting that CD9 is a partner of the Juno.

The binding of the monomeric IZUMO1 to Juno often triggers the dimerization of the IZUMO1, which in turn is said to appear in order to abolish the IZUMO1-Juno association as a Juno which becomes excluded from the adhesive surface in eth cell-oocyte assay.

CD9 and other tetraspanin are found to be able to form the higher- order protein complexes which are known as 'tetraspanin-enriched microdomains (TEMs)'. This complex is formed by direct or indirect lateral interactions with the other membrane proteins. These membrane proteins such as signaling proteins, immunoglobins, integrins, and other tetraspanins.

This further explains the reason why the tetraspanin is being involved in a range of the complex interaction. These complex interactions are like exocytosis, gamete fusion, both entry and exit of the viruses during the hot cell infection and endocytosis.

Tetraspanin is conserved in the multicellular organisms which also includes flowering plants. Tetraspanin may also found to be involved in the reproduction of the plants, as some tetraspanins are seen to accumulate in both the male and female reproductive tissues and gametes.

Among other proteins which are being reported to participate in the mammalian sperm-egg interactions are considered as the member of the ADAM (A Disintegrin and A Metalloprotease) protein family, CRISPs (Cysteine-Rich Secretory Proteins) and integrins, are the protein families that do not have true orthologs in the flowering plants.

Mouse knockout studies have often revealed the testis or sperm- specific ADAM3 (cyritestin) and ADAM2 (ferilin b)are considered to be critical for the fertilization in *in vitro* sperm migratory functions in the female reproductive tract are affected and the mature sperm is considered that it cannot bind with the egg zona pellucida (Ikawa et al., 2010).

However, the ADAM2 present in the human is a pseudogene which is suggesting that there is another ADAM protein that takes over ADAM3 function in the human sperm. Integrins are considered as an obligate (a-b) heterodimers and are considered to represent a major family of the cell surface adhesion receptors in the metazoans.

Integrin a6b1 is present on the egg and sperm and its participation in this process of the sperm-egg interaction was being supported by the fact that the Disintegrin like domains of many ADAMs are often capable of acting as an integrin ligand.

Therefore, neither a8 nor the expression of the b1 integrin subunit by the egg is considered to be essential for the female mice to be fertile. This was often suggesting that the integrins are beneficial for sperm adhesion, but it is not essential for mammalian fertilization.

The vertebrate-specific secreted CRISPs are considered as a modular protein that defines the presence of the 16 conserved cysteines. The amino -terminal module is the CAP domain (CRISP, Antigen 5, PR-1), which is found to be connected by the short hinge towards the Cysteine Rich Domain (CRD) in the carboxy- terminal.

Even though the genes which are encoding for the CRISP like proteins are not found in the plants. The CAP domain of the CRISPs is used to reveal the sequence which is similar to the plant pathogenesis- related to the protein 1 (PR-1) of the Arabidopsis, which are accumulating after the infections with the pathogens.

Transcripts that are encoding a PR-1 like protein are however, found to be abundant in the wheat egg cell and may also reveal a fertilization related function. Members of the CRISP gene family are found to be predominantly expressed in the mammalian male reproductive tract and in the venom of reptiles.

In vivo experiments are suggesting that the CRISP 1 often participates in the regulation of the signaling pathways during the process of capacitation. However, the male and the female CRIPS 1 is null in the mice which exhibited no difference in fertility, which may probably cause by the functional redundancy that is between the CRISPs.

Reptile CRISPs have been reported for regulating the range of ion channel types, and *in vitro* and *in vivo* methods that are used for identifying the mouse CRISP4 as an inhibitor which is potential for the M8 channel activity (TRPM8).

The mouse CRISP1 was recently reported that they are capable of modulating the principal of the sperm Ca^{2+}channel CatSper which is involved in the hyperactivation was found to support the idea that the CRISPs act as an endogenous protein ion channel regulator.

There is very little what is known about the gamete fusion in the flowering plants, even though the double fertilization process involves the fusion of the genetically identical make gametes (sperm cells) with the two female gametes (egg and the central cell). These reproductive cells are considered as a major distinguishing and a unique feature of the flowering plants.

Even after the third cell fusion event that was basically reported in the Arabidopsis, which occurs between the fertilized central cell and the persisting synergid cell. This is thereby preventing the attraction of the multiple pollen tubes.

The occurrence of the three-cell fusion is considered as an event which is a general mechanism and it also exists in the plants with a different ovary and ovule morphology which remains to be determined. However, the sequence of fusion events and with the precision by which the fusion of both the sperm cell takes place with the only one female gamete is being accomplished as a longstanding question in the plant reproduction research.

Using the fluorescent fusion protein, the simultaneous entry of the sperm in the internal membrane components into the egg cell and to the central celland adjustment and re-adhesion of one of the sperm cell maybe considered to require when the sperm cell pair is found to be not properly positioned between the two female gametes.

3.4.5. Karyogamy

After the plasmogamy microtubules and centrosomes, the direct migration of the gamete pronuclei for the purpose of fusion in animals (Karyogamy). Oscar Hertwig often described the process of the karyogamy in the year 1876 for the first time.

This observation is further using the very accessible sea urchin eggs and sperms, and are marking the definitive conclusion about both the egg and sperm were involved in the creation of the embryo. Moreover, there was an end of this finding formally, which was about the controversy of the spermists versus the ovists.

This means that the sperm containing the miniature individually was found to be unfolded or grew in the highly nutritive source of the egg, or that the egg can be normally be found to be sufficient for the complete development, only if the activated by the correct stimulus or a sperm.

These results are being documented that both the gametes are seemed to be involved in the developmental process. Sea urchins are being used as a paradigmatic mechanism of karyogamy- which is a sperm donated centriole (which is the egg centriole being dissembled and lost in the most animal species studied, and the rodents are significant exceptions).

These sperm-donated centrioles nucleated in the extensive microtubular arrays help in the 'capture' of the female pronucleus and draw each towards the center of the zygote, where the fusion of the two pronuclei ensues (Schatten and co-workers, 1986; Longo, F.J., and Anderson, 1968).

There are many other animals that are using a variety of variations on this theme. For example, in the ctenophore Boroe, there are multiple sperm pronuclei which enter the large, yolk-filled egg and often 'wait' to be selected.

The female pronucleus is then said to proceed towards the migrate throughout the egg cytoplasm, even after the bumping into some of the male pronuclei. At some point, it is being noticed that the female pronucleus 'choose' one of the male pronuclei and then fuses with it. They are thereby stimulating the degradation process of the remaining male pronuclei.

In some of the birds, there are multiple sperm that appear to fuse with the egg and only later one is being selected for the karyogamy. It is believed that between the extremes in the mechanism lie most of the examples of animal karyogamy.

While in the animals the gametes pronuclei approach each other and the nuclear tracking takes place along the microtubules and centrosome-dependent nuclear positioning which basically governs their movement and positioning for karyogamy.

The migration of the nuclei which is following the plasmogamy and is considered different in the flowering plants, which basically lacks the centrosomes. The two sperm cells of the flowering plants are said to migrate towards the egg and central nuclei, after then it fuses and finalizes the process of double fertilization.

Recently, it was reported that microtubules are dispensable for the migration and fusion of the male and female gamete nuclei of Arabidopsis and also for the nuclear positioning in the central cell. This process needs the presence of the intact F-actin cables.

After the process of plasmogamy, the sperm nuclei become surrounded by the aster like structure, which is made of actin and this structure migrates together with the sperm nuclei. Therefore, there are bundles of microfilaments that are being formed in a Rho-GTPase ROP8-dependent manner and it also said to appear in order to regulate the sperm nuclei migration during the process of fertilization (Kawashima et al., 2015).

The finding that helped to know that the microtubules are dispensable in the flowering plant karyogamy which is surprising since the microtubules in the animals seems to be involved at many levels. The cytoplasm which is below the surface of the plant female gametes is also found to be enriched in the microfilaments.

This may be required for the sperm cell incorporation which is similar to the animals, but they might also find to be required for the establishment of the aster like structure, which is similar to the microtubule-based aster which is associated with the animal sperm pronuclei.

In the *in vitro* fertilization experiments with isolated plant gametes, were performed in order to determine the timing between the karyogamy and plasmogamy. While from some of the experiments that were done in maize, indicated that the karyogamy occurred within the 2 to 3 hours after gamete fusion.

In the recent experiments done in rice, the movement of the sperm nuclei is adjacent to the egg nucleus and takes only 5 to 10 minutes was observed and are considered that it occurs in an act-independent manner and thus conforming the above findings in the Arabidopsis.

3.4.6. Polytubey and Prevention of Polyspermy

Eggs in the animals are considered to have a significant investment in the nutrition and resources that are basically energetically costly for a female. It is considered as an evolutionarily advantageous to have this egg cell to get fertilized by one sperm only, as the multiple sperms are fusing with the egg (which is known as polyspermy) which is usually considered as lethal to the resultant embryo.

This lethality is basically caused usually by the multiple centrioles which are being contributed by the multiple sperm that causes aberrant mitotic spindles, which is a competition for the chromosomes and cleavage abnormalities.

Therefore, the egg strategy is basically to emphasize the single sperm fusion in order to protect its investment. This is basically in contrast to the strategy that is found in the males in which the sperm have a relatively low nutrition investment in comparison to the eggs.

As a result, the male makes a large number of sperm that are being selected on the basis of their ability to get to and fuse with the egg first. The sperm onslaught is considered as advantageous for the males in the higher probability for the purpose of passing it on that male genes. Event after sometimes the killing of the egg by the polyspermy because of the highly competitive condition with the other males.

In order to be the first and outcompeting with the other males is often considered to enhance by increasing sperm production and hence the conflicts between the sexes. It is believed that the mammals with the high male competition for the mating with promiscuous females are generally considered to have larger testes for the increased sperm production.

Whereas the species with the low intraspecific male competition, in those cases the testes are smaller and the sperm production is also lower. Similarly, in the case of wind -pollinated plants, the number of sexual male organs is said to strongly increase in comparison with the species which are pollinated by the insects (Friedman, j., and Barrett, 2009).

The way by which the animal eggs counter the sperms is done by the rapid diversification of the receptors for the sperm and also by the permanent and rapid blocks towards polyspermy. The investments are being made up by the egg in order to block the polyspermy in some of the cases is considered to enormous.

For example, in the sea urchins, there are thousands of sperms that may contact the egg within the movement of each other. The egg has invested in multiple mechanisms in order to block the polyspermy. The sea urchin egg has conserved mRNA, of its organelles volume to cortical granule and of its protein. It is considered as the most conserved egg organelle that is involved in the blocking of polyspermy.

Mammalian Juno which is a binding partner of the IZUMO1 on the egg, is said to appear rapidly that is shed from the membrane into the perivitelline space after the fertilization, which may also be considered to play an essential role in the preventing the polyspermy in these species.

Animals such as insects are also using the micropyle as a limiting structure for the entry of the sperm, such that the first sperm which is entering the micropyle and that may also be sterically hindering the subsequent sperm candidates from reaching the eggs.

There are some insects egg those are even close to the micropyle which is prior to the fertilization which is as a switch between the sexual to asexual reproduction (Yashiro, T., and Matsuura, 2014).

Similar to this, the mammalian Juno, which is considered as a glycosylphosphatidylinositol- anchored protein (GAP), which is a plant -specific LORELEI (LRE). LRE is also considered to anchor to the membrane via the glycosylphosphatidylinositol- anchored protein.

LRE is being predicted as a co-receptor of the FER basically at the surface of the synergids cells and which might also be cleared by the GAP cleavage from the plasma membrane as a result of a successful pollen tube reception. It is also said to contribute towards the prevention of the entry by using the subsequent pollen tube reception into the same embryo sac. This type of condition is known as polytubey.

3.5. DOUBLE FERTILISATION

After the pollen gets deposited on the stigma, in order to reach the ovule, it must germinate and even grow through the style. It is being observed that the microspores or the pollen are containing two cells: the generative cells and the pollen tube cells.

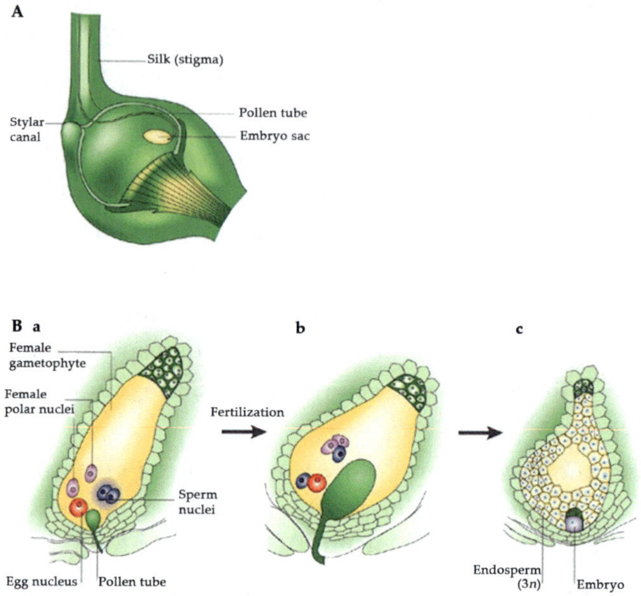

Figure 3.5: An image representing the process of double fertilization.

Source: Image by Wikimedia Commons

The pollen tube cell even grows into the pollen tube with the help of generative cell travels. The germination of the pollen tube often requires oxygen, water, and some certain types of chemical signals. As it is said to travel through the style in order to reach the embryo sac, the tissues of the style are supporting the growth of the pollen tube.

During this process, if the generative cell does not get split into the two cells then it is found that it can be now divided to form the two sperm cells. The pollen tube is being guided by the chemicals secreted by the synergids present in the embryo sac; which helps to enter the ovule sac through the micropyle.

One of the sperm from the two sperm cells fertilizes with the egg cell by forming a diploid zygote. Whereas, the other sperm cell can fuse with the two polar nuclei by simply forming a triploid cell that often develops into the endosperm.

Together with these two fertilization events which take place in the angiosperms is known as double fertilization. After the completion of the process of fertilization, no other sperm is allowed to enter. The fertilized ovule often forms the seed, whereas the tissues of the ovary are said to become the fruit which is usually enveloping the seed.

After the process of fertilization, embryonic development starts to begin. The zygote divides further divides into the form of two cells: the upper cell called a terminal cell and the lower cell called basal cell.

The division of the basal will often give rise to the suspensor, which will eventually make the connection with the maternal tissue. The suspensor often provides a route for the nutrition that is needed to be transported from the mother plant to the growing embryo.

The terminal cell also gets divided which gives rise to the globular shaped pre-embryo. In the dicot (eudicots), the developing embryo is considered to have a heart- shaped due to the presence of the two rudimentary cotyledons.

In the non-endospermic dicots, for example in *Capsella bursa*, the endosperm is developed initially, but later it is then digested. In this case, the food reserve is being moved into the two cotyledons.

These cotyledons are: embryo and the cotyledons enlarge. These two cotyledons become crowded from the inside for developing the seed and are being forced to blend. Finally, the embryo and the cotyledons fill the seed, at this point the seed is ready for the dispersal.

Embryonic development is being suspended after some time. After the seed germination, the growth of the embryo resumes. The developing seedlings will rely on the food that reserves the stored in the cotyledons until the first set of the leaves begins the process of photosynthesis.

3.5.1. Origin of Double Fertilization

The evolutionary significance of the double fertilization in the flowering plant has been questioned during the last ten years.

The original works were from the early 1900's, often suggesting that there are two karyogamy events that are occurring in a parallel manner in the female gametophyte of some of gnetales, which was reinvestigated and even confirmed by Friedman.

The rudimentary process of the double fertilization was being observed in Ephedra *nevadensis*, *Ephedra trifurca* and i. It often consists of the fusion of one sperm nucleus with the egg cell nucleus. Whereas, the second sperm nucleus is brought by the same pollen tube with the mitotic sister of the nucleus of the egg.

This process is being developed in a common ancestor of the gnetales and with that the glowering plants and one of the fusion products are seen later to be evolved into the endosperm of the angiosperms.

However, according to the recent data which separates the angiosperm from all the gymnosperms and also puts gnetales as relatives of conifers.

This basically implies that the process of double fertilization arises independently in gnetales and angiosperms. Therefore, no molecular evidences are required for determining whether the model is proposed for the origin of the double fertilization. With that, the evolutionary equivalence of the two-fertilization event is considered to be valid or if the process of double fertilization emerged differently in the angiosperms.

3.5.2. Structural and Cytological Perspectives on Double Fertilization

It is considered that the years following the discovery of the double fertilization in the angiosperms are witnessed not only as a steady increase in the number of the species which are showing this phenomenon.

It is also considered to attempt to link it with the other aspects of the sexual reproduction in the flowering plants and are being established as a model system used to study cell biology and nuclear cytology of the fusion.

The identification of the discrete phases is being done in the double fertilization episodes such as the arrival of the pollen tube in the embryo sac, the entry of the sperm into the egg cell or to the central cell, release of the sperm into the embryo sac, alignment and migration of the sperm nuclei with the female nuclei and nuclear fusion (Russell, 1992).

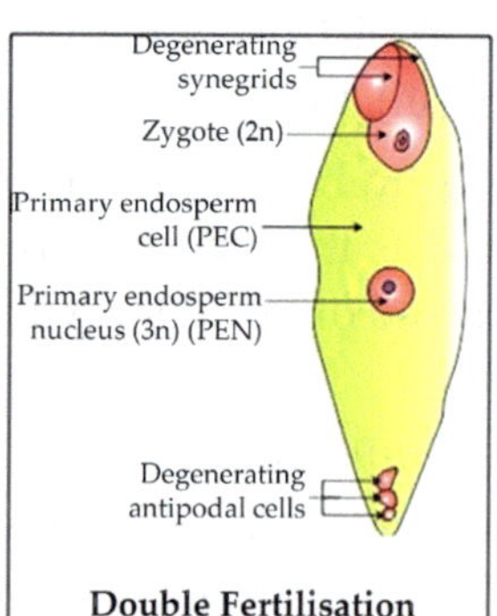

Figure 3.6: Structure of double fertilization.

Source: Image by Topper Learning

The uncovering of the details that are linked to the nuclear fusion is said to occur in the shrouded environment of the embryo sac in order to have herald double fertilization process that has been daunted plant embryologist for a quite long time. But there are some inroads that have been made towards this goal.

Faure (2001) and Faure and Dumas (2001) were the first who addressed the main issues that in the cell biology of the double fertilization, whereas Russell (1992) has reviewed the most critical ultra-structural aspect of this process.

3.5.3. Universality of Double Fertilization in Angiosperms

The momentum which was created in the waning years of the nineteenth century in order to establish the double fertilization as a unique feature in the reproductive biology of the angiosperms. This was basically followed by a sustained effort that was done in the first 2 years of the twentieth century.

This was leading towards the discovery of this phenomenon in the addition of the members : Ranunculaceae (Guignard, 1901c), Juglandaceae (Karsten, 1902), Liliaceae (Ikeda, 1902), Monotropaceae (Shibata, 1902).

This discovery was also members of the plants who are belonging to Najadaceae (Guignard, 1901b), Asclepiadaceae (Frye, 1902), Ceratophyllaceae (Strasburger, 1902), Poaceae (Guignard, 1901a), Gentianaceae (Guignard, 1901d), Solanaceae, and Brassicaceae (Guignard, 1902).

Guerin (1904) in a monograph devoted entirely on the topic of fertilization in the seed plants and Coulter and Chamberlain (1912) in their classic book which was on the Morphology of Angiosperms, basically refers to the 16 families of the angiosperms. It also encompasses about 40 of the genera and about 60 species those are definitely known to have a second fertilization event.

From that time, along with the presence of the reduced female gametophyte and embryo nourishing endosperm, it was believed that the occurrence of the double fertilization was being accepted as a regular feature that was associated with the reproductive biology of the angiosperm.

Under this assumption, there were only the infrequent references that are used for double fertilization in the numerous publications from the mid of the 1900s to the present. This dealt with the variability and diversity of the reproductive processes in the flowering plants containing special reference to their embryogenesis and endosperm development (Johansen, 1950; Maheshwari, 1950; Davis, 1966; Johri et al., 1992).

3.6. TRIPLOIDY OF THE ENDOSPERM

Apart from those exceptions the production of the endosperm with the high ploidy, the widespread occurrence of the triploid endosperm that was among the angiosperms with the 2:1 maternal: paternal genome ratio, raises the question related to the evolutionary significance of the asymmetric parental genome balance.

In the year 1989, Haig and Westoby formulated that the parental conflict theory which states that the biparental contribution towards the endosperm reflects that the selfish interests of the father and mother over the resource allocation to the progeny.

With the access of the maternal chromosomes which was over the paternal ones, the mother can exert greater control over the nutrient allocation to the progeny in relation to own fitness. At the molecular level, the parent-of-origin specific mechanisms that are used for the expression of the selfish parental interest that basically act via cytoplasmic deposited determinants.

It also acts as dosage- sensitive gene products or genomic imprinting. Because of its role in the embryo nourishment, it is believed that the endosperm appears to be the battleground for such conflicts, whereas, in the evolutionary context, the endosperm is being viewed as the angiosperm homolog which is equivalent to the gymnosperm female gametophyte.

But it is considered to provide the fitness advantages at the organismal level in order to thanks the acquisition of the paternal contribution, known as double fertilization. According to the current phylogenetic distribution of the triploid and diploid endosperm, it does not indicate that which was the basal state, but it demonstrates that either the diploid or triploid of the endosperm was gained or lost twice among the flowering plants.

3.7. CONCLUSION

Gamete fusion is considered as a necessary pre-requisite for the process of sexual reproduction that takes place in all the organisms. One might also think about the underlying mechanism are also highly conserved. It is made clear that the organisms have been evolved enormously from the diversity in order to accomplish the same goal.

It often appears as a unique reproductive method that requisite in order to have a successful species competition and to allow the organism to test the new dimensions within which they can occupy their niche more effectively.

With the help of exploration of different organisms, the researchers have finally found about how the process of fertilization takes place in a diversified organism.

With the focus on the process of fertilization that takes place in the flowering plants, this chapter delves into the process of fertilization that takes place in the plant in different steps. It also explains the molecular machinery that is involved in the gamete interaction in plants and animals.

It is observed that the structure of micropyle is found in both the plants and the animal organisms that basically restrict the entry site of the sperm to the egg. There are also other processes like the loss of the sperm motility and the usage of the actin cytoskeleton instead of using the microtubules in order to achieve gamete nuclei fusion in the plants.

Flowers are basically considered as a seat of sexual reproduction that takes place in the angiosperms. In the flowering plants, there is the presence of stamens that basically refers to the male reproductive organ and gynoecium is considered to consist of pistils that represent the female reproductive organs.

An anther is considered as typical bilobed, dithecous and tetrasporangiate. Pollen grains often develop inside the microsporangia. The microsporangium is found to be surrounded by the four wall layers, the epidermis, endothecium, middle layer and the tapetum.

It is considered that the process of fertilization takes place in three ways known as Porogamy, Chalazogamy and Mesogamy. These three types are different and constitute the process of fertilization by using different aspects.

During the of fertilization, there is another case when the female gamete joins the two sperm or male gametes and forms the polyploid. This process is known as double fertilization. This process takes place in a few of the flowering plants.

REFERENCES

1. 2015. *SEXUAL REPRODUCTION IN FLOWERING PLANTS*. [ebook] Available at: <http://ncert.nic.in/NCERTS/l/lebo102.pdf> [Accessed 18 May 2020].
2. Baroux, C., Spillane, C. and Grossniklaus, U., 2002. *Evolutionary Origins Of The Endosperm In Flowering Plants*. [online] genome biology biomedcentral. Available at: <https://genomebiology.biomedcentral.com/articles/10.1186/gb-2002-3-9-reviews1026> [Accessed 18 May 2020].
3. Bio.libretexts.org. n.d. *32.2D: Double Fertilization In Plants - Biology Libretexts*. [online] Available at: <https://bio.libretexts.org/Bookshelves/Introductory_and_General_Biology/Book%3A_General_Biology_(Boundless)/32%3A_Plant_Reproductive_Development_and_Structure/32.2%3A_Pollination_and_Fertilization/32.2D%3A_Double_Fertilization_in_Plants> [Accessed 18 May 2020].
4. BYJUS. n.d. *Fertilization Of Plants, Process Of Fertilization In Plants | Byju's*. [online] Available at: <https://byjus.com/biology/fertilization-in-plants/> [Accessed 18 May 2020].
5. Dresselhaus, T., Sprunck, S. and Wessel, G., 2016. Fertilization Mechanisms in Flowering Plants. *Current Biology*, [online] 26(3), pp.R125-R139. Available at: <https://www.researchgate.net/publication/293644689_Fertilization_Mechanisms_in_Flowering_Plants> [Accessed 18 May 2020].
6. Dresselhaus, T., Sprunck, S. and Wessel, G., 2016. Fertilization Mechanisms in Flowering Plants. *Current Biology*, [online] 26(3), pp.R125-R139. Available at: <https://www.ncbi.nlm.nih.gov/pmc/articles/PMC4934421/> [Accessed 18 May 2020].
7. Faure, J. and Dumas, C., 2001. *Fertilization In Flowering Plants. New Approaches For An Old Story*. [ebook] www.plantphysiol.org. Available at: <http://www.plantphysiol.org/content/plantphysiol/125/1/102.full.pdf> [Accessed 18 May 2020].
8. Faure, J., 1999. *Double Fertilization In Flowering Plants: Origin, Mechanisms And New Information From In Vitro Fertilization*. [online] SpringerLink. Available at: <https://link.springer.com/chapter/10.1007/978-3-642-59969-9_7> [Accessed 18 May 2020].
9. Faure, J., 2001. Double fertilization in flowering plants: Discovery,

study methods and mechanisms. *Comptes Rendus de l'Académie des Sciences - Series III - Sciences de la Vie*, [online] 324(6), pp.551-558. Available at: <https://www.researchgate.net/publication/11885752_Double_fertilization_in_flowering_plants_Discovery_study_methods_and_mechanisms> [Accessed 18 May 2020].

10. Faure, J., 2001. Double fertilization in flowering plants: Discovery, study methods and mechanisms. *Comptes Rendus de l'Académie des Sciences - Series III - Sciences de la Vie*, [online] 324(6), pp.551-558. Available at: <https://www.sciencedirect.com/science/article/abs/pii/S0764446901013257?via%3Dihub> [Accessed 18 May 2020].
11. Raghavan, V., 2003. Some reflections on double fertilization, from its discovery to the present. *New Phytologist*, [online] 159(3), pp.565-583. Available at: <https://nph.onlinelibrary.wiley.com/doi/full/10.1046/j.1469-8137.2003.00846.x> [Accessed 18 May 2020].
12. Shivanna, K., 2016. *Fertilization In Flowering Plants*. [ebook] RESONANCE. Available at: <https://www.ias.ac.in/article/fulltext/reso/021/11/1007-1018> [Accessed 18 May 2020].
13. Sprunck, S. and Dresselhaus, T., 2015. Three Cell Fusions during Double Fertilization. *Cell*, [online] 161(4), pp.708-709. Available at: <https://core.ac.uk/download/pdf/81998227.pdf> [Accessed 18 May 2020].
14. Bleckmann et al., 2014
15. Coulter and Chamberlain, 1912
16. Davis, 1966
17. Denninger and co-workers, 2014
18. Dresselhaus et al., 2009
19. Drews et al. 2011
20. Evans, J.P., 2012
21. Faure and Dumas, 2001
22. Faure, 2001
23. Friedman, j., and Barrett, 2009
24. Frye, 1902
25. Guerin, 1904
26. Guignard, 1902
27. Haig and Westoby, 1989

28. Higashiyama et al., 2015
29. Ikawa et al., 2010
30. Ikeda, 1902
31. Johansen, 1950
32. Johri et al., 1992
33. Karsten, 1902
34. Kawashima et al., 2015
35. Leshem et al., 2013
36. Longo, F.J., and Anderson, 1968
37. Maheshwari, 1950
38. Russell, 1992
39. Schatten and co-workers, 1986
40. Shibata, 1902
41. Spunk et al.,2011
42. Strasburger, 1902
43. Yashiro, T., and Matsuura, 2014

Chapter 4

Gametogenesis

CONTENTS

This chapter named as 'Gametogenesis' basically refers to the process of formation and development of gamete whether male or female in the plants. In the first half of the chapter, it provides information regarding the basic concept of gamete development and this topic includes an explanation of the regulation of sporocyte development, sporophytic and gametophytic interactions, meiotic development, chromosome pairing and synapsis and then transcriptional regulation of gametophytic development. In the second half, the readers will get insights about the development of female gametophyte. Then it provides knowledge about the make gametophyte and its development in angiosperms.

4.1. INTRODUCTION

In higher plants, the process of gamete development mirrors the process which can be seen in animals. However, despite the fact that there is conservation of process and substantial information can further be obtained by the comparative analysis of distantly related organisms, there still remain distinct differences, which are particular to the organisms which are concerned.

Such kinds of differences include germ-line specifications. This means that the plants have the ability of producing gametes at a late developmental stage which has not been pre-determined. On the other hand, in animals, the gamete formation is confined to distinct regions which are established at an early embryonic stage.

In addition to this, despite the fact that fertility can be frequently compromised by the mutation of genes related to the gamete formation having defects frequently resulting in decreased fertility, they are rarely considered lethal in plants.

These defects can be maintained most of the time since many plants produce both male as well as female gametes and often, they tend to form one type of gamete that remains unaffected by the mutation. Also, the plants are able to reproduce asexually or even through non-fertilizing means; an example of apomixes can be considered.

The key stage, while developing the process for higher plants, is the initiation of meiosis and the reason is that it provides the trigger in order to split the sporophytic as well as gametophytic generations.

Despite this, much information is known regarding the process of meiosis. There has been a detection of a number of homologies between animals and plants about which little is known related to the plant signals which tend to switch development from mitosis to meiosis.

Male as well as female gamete production is similar in higher plants having a number of genes playing a role while developing both types of gametes. Nonetheless, the subsequent fate of the male as well as female meiotic products is distinct.

Considering the male gamete production, all of the four meiotic products that are microspores (which refer to the male spore which will give rise to the make gametophyte) tend to develop in order to produce viable and fully functional pollen grains (McCormick 2004, Scott et al. 2004).

Nevertheless, considering the female gamete production, usually, only one of the four megaspores (which refer to the female spores which will give rise to the female gametophyte) will continue to develop, while the remaining meiotic products tend to degenerate by a form of programmed cell death (Schneitz et al. 1998, Yadegari & Drews 2004).

Related to this, much is known about gamete formation in higher plants and this information has been obtained while studying the model dicotyledonous plant *Arabidopsis thaliana* (Meinke et al. 1998) and the monocotyledon maize.

4.2. GAMETE DEVELOPMENT

Later in plant development, the development of reproductive organs and their meiocytes, not like animals in which primordial germ cell development tends to occur early during embryonic development (Wylie 1999, Zhao & Garbers 2002).

In higher plants, the formation of gametes occurs within defined organs such as the anther and ovule. The anther in the male contains the haploid pollen and the ovule in the female contains the embryo sac and so, is the progenitor of the seed.

The initiation process tends to occur in the same manner considering that both male as well as female gametes having the differentiation of archesporial cells and initiation of meiosis.

The initiation of male gamete formation takes place by periclinal divisions (which is parallel to the anther wall) in a number of adjacent hypodermal cells within the anther primordium. These tend to develop into archespores: typically four located at the four corners of each and every anther primordium.

After that, the archesporial cells divide mitotically in order to form the inner primary sporogenous cells, which later on go for the formation of PMCs that is Pollen Mother Cells and primary parietal cells (Scott et al. 2004).

Considering the case of Arabidopsis, it can be said that there are around 30 PMCs per anther. The primary parietal cells go through further mitotic divisions for the formation of the anther wall, which contains the endothecium, middle cell layer and tapetum.

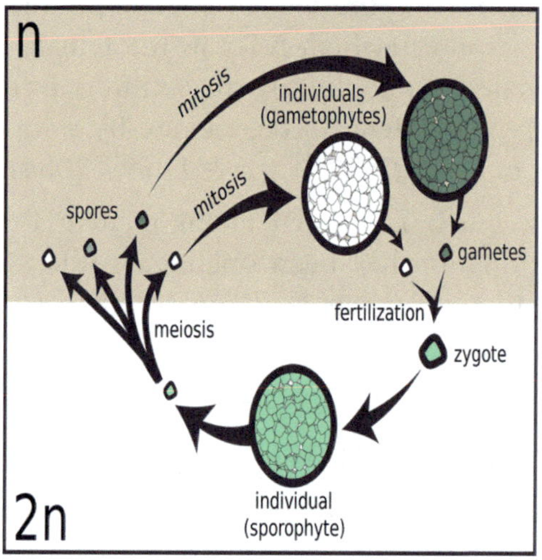

Figure 4.1: Sporic Meiosis.

Source: Image by Wikimedia Commons

The tapetum tends to surround the sporogenous cells and moreover, serves for nourishing the developing microspores during the process of PMC meiosis as well as microspore maturation. Callose (b1,3 glucan) tends to form around the PMCs and later on, between the meiotic products, the normal plasmodesmata connections in between the microspores are severed.

However, the plasmodesmata expand in order to form 'cytomictic channels' of up to 0.5mm due to which possibly some cytoplasmic exchange could occur during the process (Heslop-Harrison 1966).

After the process of meiosis, the breakdown of callose takes place by the release of a b1,3 glucanase (callase) by the tapetum. It is important to consider the timing of this for the purpose of pollen viability which has been used as an approach so as to engineer male sterility in an artificial manner (Worrall et al. 1992, Rhee & Somerville. 1998).

Nevertheless, the analysis of such lines provides suggestions that the callose wall is not required for male meiosis (Scott et al. 2004).

After that, the microspores go through a process of wall deposition, which is co-ordinated by the tapetum but also in part, by the microspores or immature pollen grains (Piffanelli et al. 1998, Scott et al. 2004).

Furthermore, two mitotic divisions tend to occur for producing functional male gametes. These tend to occur during the late stages of pollen development which is seen in the species that tend to produce tricellular pollen such as Arabidopsis.

In an alternative manner, the first mitotic division (PMI that is pollen mitosis I) can occur in the pollen grain. So that, while releasing, the mature pollen is bicellular, such as tobacco, and moreover, further mitotic division (PMII that is pollen mitosis II) tends to occur during pollen tube development.

The PMI division is said to be asymmetric and further, it establishes polarity as well as determinacy within the pollen grain (Eady et al. 1995).

Furthermore, two cells are formed which tend to have very distinct fates, the larger vegetative cell tends to accumulate a dense cytoplasm which contains lipids, proteins and also, carbohydrates (McCormick 2004).

Further, this does not divide, but it tends to provide the storage compounds for the active period of growth that is related to the production of a pollen tube. There is another generative cell which is smaller having highly condensed chromatin and subsequently, it goes via PMII in order to produce two sperm cells.

Considering the case of plants, double fertilization is characteristically seen (Faure 2001), along with one sperm cell fusing with the egg cell in order to form an embryo, whereas, the other fuses with the diploid central cell in order to form the endosperm.

The endosperm is said to be a triploid tissue which tends to serve in order to nourish the developing embryo before the commencement of functional photosynthesis by the developing seedling.

The formation of the female gamete occurs within the ovule (Schneitz et al. 1998, Yang & Sundaresan 2000, Skinner et al. 2004). There are different types of female gametophyte development that have been described;

nevertheless, more than 70 % of flowering plants tend to exhibit the Polygonium type pattern of development which results in an eight nucleate embryo sac (Drews & Yadegari 2002, Yadegari & Drews 2004).

Typically, an Arabidopsis gynoecium contains approximately 50 ovules. Each and every ovule contains a nucellus which tends to enclose the embryo sac, and this is encased within two integuments that are the progenitors of the seed coat.

The growth of the integuments takes place asymmetrically which can result in the curvature which can typically be observed in mature ovules (Schneitz et al. 1998, Skinner et al. 2004); the funiculus or stalk that connects the ovule to the placenta.

The nucellus consists of the spore - bearing tissue (that is MMC which means megaspore mother cell). The MMC goes through meiosis in order to form four haploid megaspores. However, this takes place in one and three of these degenerate through programmed cell death.

This usually leaves only one functional megaspore at the chalazal end. Then, this goes through four successive mitotic divisions and this is done in order to form the eight - nucleate embryo sac (Schneitz et al. 1998, Skinner et al. 2004).

Among these nuclei, two tend to fuse in order to form the central cell, which is diploid and is fertilized later on by one of the sperm cells for the formation of the triploid endosperm. Therefore, a typical embryo sac contains seven cells with eight nuclei.

Usually, the nucellus which is encapsulated by two integuments, the exception is for a small cleft at the base, termed the micropyle. Through this, the pollen tube grows in order to allow double fertilization by the two sperm cells.

The nucellar tissue undergoes degeneration during the development of the embryo sac. Eventually, the integuments form the seed coat and these are also thought to have a role in the regulation of the nucellar degeneration and sporophyte development (Schneitz et al. 1998, Skinner et al. 2004).

4.2.1. Regulation of Sporocyte Development

In higher plants, the meiosis proceeds in the same manner to other higher eukaryotes, having a number of processes including pairing and recombination conserved. Nevertheless, there is not much known about the critical signals.

These tend to switch cellular development from a diploid, sporophytic generation to a haploid, generative generation. This switch tends to occur late in the development of plant cells, and these can be induced in the tissue which is previously undifferentiated and furthermore, divides the sporophytic as well as gametophytic generations.

A number of genes function while determining the specification of the male or female germline and identification of these genes has been done in *Arabidopsis*. In wild-type plants and meiotic mutants, detailed analysis of the developmental process of meiosis has been conducted (Ross *et al.* 1996, 1997, Bhatt *et al.* 2001).

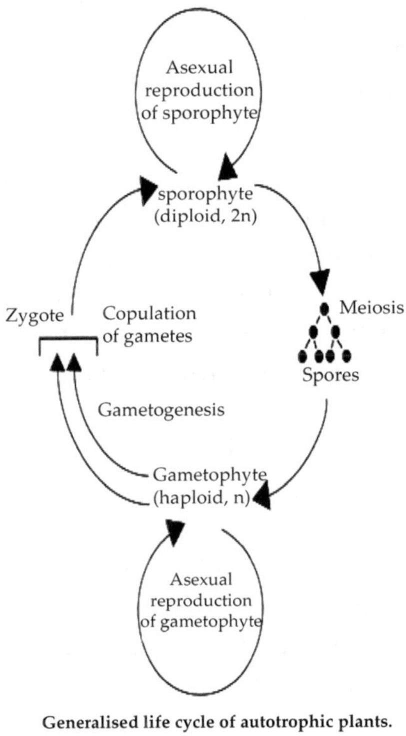

Figure 4.2: Generalised lifecycle of autotrophic plants.

Source: Image by Wikimedia Commons

The *SPORO-CYTELESS (SPL)/NOZZLE (NZZ)* **gene has a homology to MADS box transcriptional factors. This is required in order to initiate**

the sporogenesis in both male as well as female organs (Schiefthaler *et al.* 1999, Yang *et al.* 1999*b*). In the *spl/nzz* **mutant, archesporial initiation tends to occur in a normal manner. Nevertheless, subsequent differentiation of both the male as well as female sporocytes is blocked.**

Along with this, the occurrence of the anther wall development fails, but integument development is normal despite the fact that the nucellus development is delayed (Yang *et al.* 1999*b*). This provides suggestions about the interactions of *SPL/NZZ* **with the surrounding cells and these suggestions tend to determine the outcome within the anther and ovule.**

> Yang *et al.* (1999*b*) proposed that the microsporocytes promote, through cell – to – cell communication or through the passage of a signal, the differentiation as well as growth of the anther parietal layers. Therefore, the cell wall development takes place.

Considering the case of the *sterile apetela (sap)* mutant, the megasporocytes fail to complete meiosis and there are effects on the floral number and organ number also (Byzova *et al.* 1999). However, it is proposed that *SAP* may act downstream of *SPL/NZZ*, and the reason is that the megasporocyte formation occurs in *sap* **mutants.**

The MAC1 that is *MULTIPLE ARCHESPORIAL CELLS 1* **gene from maize tends to appear in order to function at an earlier stage to** *SPL/NZZ* because it does alterations of the specification of archesporial cells.

In the *mac1* **mutant, a number of various archesporial cells** form in the ovule (Sheridan *et al.* 1996); nevertheless, in the anther, the archesporial specification is considered to be unaffected but meiotic arrest is seen.

This further implies that *SPL/NZZ* **can even act upstream of** *MAC1* **in the anthers** (Yang *et al.* 1999*b*). There can be a similar phenotype of altered numbers of archesporial cells which is seen in the *Arabidopsis* **mutants** *extra sporogenous cells* (*exs*)/*excess microsporocytes1 (ems1)* **and** *tapetal determinant 1 (tpdl)* **and the** *EXS/EMS1* **rice orthologue MSP1 that is** *MULTIPLE SPOROCYTE1* (Nonomura *et al.* 2003).

EXS/EMS1 **tends to appear in order to regulate the number of cells which further divide in the L2 layer of the anther (goes so as to form archesporial cells**) and so, the number of sporogenous cell initials.

In the *exs/ems1* **mutant, the formation of the additional meiocytes takes place, but the tapetal and middle cell layers are also missing. This can also be seen in the** *gus negative1* (*gne1*) mutant (Sorensen *et al.* 2002).

The *EXS/EMS1* **gene tends to encode for a putative serine/threonine LRR which means Leucine- Rich Repeat receptor kinase** (Canales et al. 2002, Zhao et al. 2002).

Furthermore, this is considered to act by complexing with a number of various partners in different tissues. Therefore it further restricts growth within the L2 layer of the anther or even enhances growth in the embryo (Canales *et al.* 2002).

There can be a similar phenotype of the tapetal cells which is being replaced by sporogenous cells and this is observed in the mutant *tpd1. TPD1* **is expressed in a predominant manner within the microsporocytes which can further reflect interactions taking place in between the developing sporocytes and tapetal cells, along with the formation of sporocytes as a default pathway in the absence of** *TPD1*(Yang *et al.* 2003*b*).

TPD1 **tends to encode for a small protein of 176 amino acids and therefore, it considered acting in coordination with** *EXS/EMS1* (Yang *et al.* 2003*b*). There can be the formation of sporogenous cells in the absence of the tapetum in *tpd1* **and** *exs/ems1*mutants which in turn implies that sporocytes can form, and also, that the initiation of the process of meiosis can be done in the absence of the tapetal cell layer.

Nevertheless, both mutants are male sterile along with degeneration of the microsporocytes which occur during the process of meiosis II. This in turn suggests that the development of viable pollen requires a functional tapetum.

From sporophytic to gametophytic development, the *Arabidopsis* **loci** *SWITCH1/DYAD* **and the maize** *AMEIOTIC1* **have been implicated in the switch.**

Considering the *ameiotic* **mutant, the formation of both mega as well as microsporocytes are done, but meiosis does not proceed, along with** *ameiotic1* **meiocytes which further carries out a form of division that is mitotic-like instead of meiosis after which they tend to degenerate** (Golubovskaya *et al.* 1993, 1997).

Considering the ovules of the *swi1* **mutant, the archesporial cells** tend to undergo a further mitotic division prior to entering meiosis (Siddiqi *et al.* 2000).

There have a number of genes that have been identified by muta-genesis as having a clear phenotype in relation to meiosis and the examples include *SWI1, SYN1, ASY1.* These are also expressed within the vegetative tissues.

Nevertheless, there has been no mitotic phenotype which has been seen along with these genes. This further implies that they may be non-essential for mitosis, that there may be functional redundancy or that some form of post-translational regulation can regulate the developmental role of such type of proteins (Mercier *et al.* 2003).

4.2.2. Sporophytic and Gametophytic Interactions

There has been evidence that indicates that the meiosis and the subsequent development of the gametes is regulated by the developing gametophytes in part, and along with this, this is also done by the surrounding sporophytic tissue.

This is true in both the cases for male as well as female gametes. In these, the tapetal tissues tend to play a major role while developing the pollen and the nucellus for egg cells.

The tapetum is said to be a transient as well as metabolically active sporophytic cell layer. This tends to enclose the sporogenous tissue within the anther locule (Chapman 1987). Moreover, it plays a critical role in the nutrition of microspores, while regulating sporogenesis and in the development of the pollen wall.

The tapetal tissue is considered to be highly metabolically active and tends to undergo endo-replication in order to form bi- or multinucleate cells, along with a proliferation of ribosomes (Weiss & Maluszynska 2001).

Polyploidisation of the tapetum is said to occur for increasing the potential for the purpose of synthesis without cell division. As per the observation done in cytoplasmic male sterile lines, the process of microspogenesis is extremely sensitive to mitochondrial mutations (Schnable & Wise 1998).

These in turn, typically show tapetal aberrations but these do not have specific vegetative phenotypes. This suggests that the microsporogenesis as well as tapetal development, is an extremely energy-demanding process.

The tapetum tends to play a major secretory role in sporogenesis, in direct analogy so as to nurse cells in the mammalian systems. A key role is played by tapetum in the process of secretion of pollen wall materials, primarily sporopollenin, which contains mixed polymers of fatty acid and phenylpropanoid derivatives.

Furthermore, it provides a critical function in pollen dispersal as well as pollen–stigma interactions (Piffanelli *et al.* 1998, Aouali *et al.* 2001).

In addition to this, the tapetum is said to be critical in the secretion of callase (β1,3 glucanase) for regulating the breakdown of the callose (β1,3-linked glucan) wall surrounding the tetrads.

During the process of microspore maturation, the tapetum tends to undergo a process of highly regulated programmed cell death and this results in the release of the cell wall ,as well as lipid compounds, and its final disintegration before anthesis.

The pattern which is must similar to the nucellus degradation is observed during the later stages of female embryo sac development. However, the occurrence of normal breakdown does not take place in the *sap* **mutant in which megasporogenesis** is arrested (Byzova *et al.* 1999).

Therefore, it is speculated that a functional megaspore can be required for nucellus degradation and such signals from the integument are required in order to regulate the progression of meiosis in the female gametes (Yang & Sundaresan. 2000).

4.2.3. Meiotic Development

Meiosis is considered to be a critical as well as, highly conserved stage which occurs during the sexual reproduction of all eukaryotes.

A single round of DNA replication is combined along with two successive divisions which in turn, forms four haploid products. These haploid products allow the process of reassortment as well as segregation of genetic information.

When the first separation occurs then it involves the segregation of homologous chromosomes, which have previously undergone pairing in order to form bivalents, synapsis and recombination. Then the second division occurs which involves the separation of sister chromatids of each chromosome and this further results in four haploid gametes.

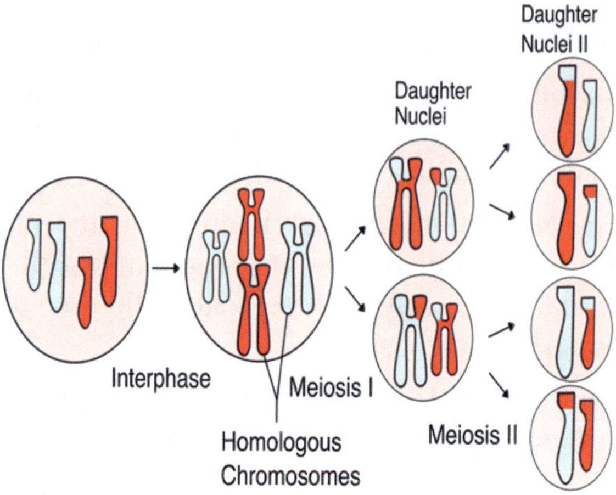

Figure 4.3: Meiosis.

Source: Image by Wikipedia

Meiosis tends to appear in order to be highly regulated and also, for proceeding in a synchronous manner. Nevertheless, as seen with a number of angiosperms, there is a high degree of asynchrony that can be found between male and female meiosis. In *Arabidopsis*, the male meiosis is completed before female meiosis reaches prophase I (Caryl *et al.* 2003).

There has been evidence of large numbers of genes that have their established roles in meiosis and mitosis in yeast as well as humans. Despite the fact that there have been sequence analyses which has been used for the identification of homologs and aid functional classification of genes, it has had mixed success.

For instance, while the synaptonemal complex (SC) exhibits high levels of functional similarity in plants and animals, the search for functional homologs of SC proteins has been much unsuccessful. Nevertheless, to some extent, we can expectat because of the fact that there are low levels of homology which are observed between yeast and mammalian systems.

Let's take an example. The yeast *ZAPI1* **and rat** *SCP1* **genes both tend to encode for proteins that are related to the synaptonemal complex. However, the only regions of similarity are within the coiled-coil regions** (Heyting 1996).

There can be a more successful approach for identifying the genes having a critical role in plant meiosis which has been the analysis of sterile or reduced fertility mutants (Caryl *et al.* 2003).

The identification of a number of genes has been done which act in either male or female meiosis, or even in both, and the comparison of their roles have subsequently been done with those in yeast and higher eukaryotes.

4.2.4. Chromosome Pairing and Synapsis

There are five distinct as well as conserved stages, during which the process of chromosome condensation and pairing occurs during meiosis 1 (Ross *et al.* 1996, 1997).

During leptotene, the condensation of the chromosomes takes place, and this tends to appear as thin thread-like structures and further, the homologous pairing is initiated. During zygotene, the pairing continues until full synapsis is seen in pachytene.

This is the point at which the homologs have fully paired and further, appear as thick structures and the formation of SC takes place. During diplotene, the occurrence of desynapsis starts along with the homologs remaining attached at the points of chiasmata.

By diakinesis, the chromosomes tend to appear as condensed bivalents and the homologs are critically attached as they enter anaphase I. During late leptotene/zygotene, the occurrence of meiotic recombination takes place.

Cohesins are considered to be a evolutionarily conserved proteins, the implication of which has been done in the close association seen between replicated sister chromatids during the process of prophase I. Furthermore, they have a potential role in recombination and double - strand break repair (Nasmyth 2001). In the process of budding and fission yeast, two cohesins have been characterised.

First and foremost, the identification of RAD21 protein was done from *Schizosaccharomyces pombe* and considered as essential in double - stranded DNA repair during the process of mitotic growth.

RAD21 **is said to be a cell cycle - regulated phosphoprotein required for the process of chromosome** condensation and sister chromatid cohesion. There has been an identification of a similar meiotically expressed gene, *RAD8* in *S. pombe*.

- The identification of a small *RAD21* **family al**ong with cohesin activity was also done in *Arabidopsis*. There have been predictions that these can have defined roles which are related to the chromosome condensation during the process of mitosis and meiosis (Bai *et al.* 1999).

The *DIF1/SYN1* **gene has the greatest sequence homology to** *RAD21*; however, *DIF1/SYN1* **is said to be meiotically active, which exhibits both male as well as female sterility, when the mutation is done.**

This happens because of the lack of chromosome condensation and pairing during the leptoneme of meiosis I (Bai *et al.* 1999, Bhatt *et al.* 1999).

Mutation of the *Arabidopsis SOLO DANCERS* (*SDS*) gene results in a severe defect in homolog synapsis, recombination and bivalent formation during prophase I of meiosis in both male and female sporocytes. This in turn, causes random distribution of the chromosomes and the abnormal meiotic products (Azumi *et al.* 2002).

In the C-terminal region, SDS consists of a conserved domain that is cyclin-like. This domain has the most similarity to the B2 subtype of *Arabidopsis* **B cyclins.**

Nevertheless, the levels of similarity are considered to be much similar to that which can be seen between the *Arabidopsis* **A, B and D types of cyclins. Therefore, it is speculated that** *SDS* **can represent a new type of cyclin** (Azumi *et al.* 2002).

It is proposed that a novel meiosis-specific cyclin is required in homologue interactions which ,in turn, suggested that a cyclin dependent kinase can play a key role in controlling chromosomal events during the process of meiotic prophase I. *SDS* **appears to be meiosis- specific which cannot be expressed in vegetative tissue** (Azumi *et al.* 2002).

Arabidopsis **SWITCH (SWI1) is said to be a novel protein which tends to appear in order to colocalize with DNA and also, it is required for the promotion of the sister chromatid cohesion, axial element formation and recombination. In the** *swi1* **mutant, it has been observed that there are 20 individual chromatids at metaphase 1 instead of five bivalents**.

However, the condensation is seen in the *swi1* **mutant further implying that SWI1 is involved while initiating the cohesion instead of condensation. Along with this, it appears not to be involved in the maintenance of cohesion, since the protein does not persist beyond meiosis I.** (Mercier *et al.* 2003).

SDS is said to be involved in the timing of chromosome separation by the phosphorylation of proteins which are involved in sister chromosome cohesion. Therefore, it is possible that SDS can regulate SWI1 activity (Azumi *et al.* 2002).

The *Arabidopsis ASY1* **gene is important for both the cases male as well as female homologous chromosome** synapsis and also, it has been connected to the SC. When the process of mutation is done, then the chromosomes tend to align during diakinesis, but no bivalents or chiasmata are detected. There has been the identification of a second *ASY2* **gene also in the** *Arabidopsis* **genome but the function of this is still not known.**

In addition to this, the *ASY1*gene transcript is detected in vegetative tissue but there is no vegetative phenotype observed.

There has been expression in vegetative tissue which is also seen with a number of other meiotic genes although, there is a lack of mitotic phenotype, for example, *syn1* (Bai *et al.* 1999). The detection of ASY1 is done in PMCs during the process of meiotic interphase as punctate foci which are distributed over the chromatin.

As leptotene progresses, it tends to relate to the axial loops, but it is not the associated extended chromatin loops and furthermore, it tends to extend over the length of the chromatin. In this, it does not include the telomeres (Armstrong *et al.* 2002).

As the homologs desynapse, the protein tends to disappear. The ASY1 protein has homology to the N-terminus of the yeast homeodomain protein 1 (HOP1) axial core-associated protein (Caryl *et al.* 2000).

The region of homology consists of a HORMA domain which is found in proteins that is related to the chromatin. Over the remainder of the protein, homology is however low. Therefore functional similarity between the proteins cannot be extended beyond the ability of both proteins in order to bind chromatin (Armstrong *et al.* 2002).

HOP1 is observed to be in relation with a similar manner to ASY1 as a series of foci in leptotene, which is dependent upon another protein RED1 in yeast. This protein is needed for axial element assembly (Smith & Roeder 1997). Both *ASY1* **as well as** *RED1* **have similar expression patterns and furthermore, appear to work together albeit along with defined roles.**

Nevertheless, there is no sequence homology which is observed in between RED1 and ASY1, or to any other *Arabidopsis* **proteins.**

Therefore, the ASY1 can serve a similar role to the HOP1/RED1 proteins having an important function related to the interface between the SC and axis-associated chromatin (Armstrong *et al.* 2002).

Homologs of a number of genes are required for the catalytic steps of recombination in yeast and the vertebrates. These are now been identified in *Arabidopsis*, which include *RAD51, SPO11, DMC1, RAD50, MRE11* (Klimyuk & Jones 1997, Doutriaux *et al.* 1998, Couteau *et al.* 1999, Gallego *et al.* 2001, Gallego & White 2001, Grelon *et al.* 2001, Puizina *et al.* 2004).

The yeast DMC1 protein has been identified and considered to be important in order to form bivalents and partitioning of chromosomes during meiosis I.

When the mutation of the *Arabidopsis* **orthologue,** *AtDMC1* takes place then it tends to result in a reduction of fertility to less than 1.5 percent and a failure of chromosome partitioning, despite the fact that, there is no significant occurrence of a complete meiotic arrest as seen in yeast (Couteau *et al.* 1999).

This provides suggestions that a distinct mechanism of checkpoint activation or progression surveillance can also exist in plants when it is compared with yeast (Bhatt *et al.* 1999).

The Mre11 protein complex contains Mre11,Rad50 and Xrs2 and also, this has been characterized in *Saccharomyces cerevisiae* **and vertebrate systems. This tends to play a central role in the initial stages of** DSB: double - strand break repair (Ray & Langer 2002).

This system of DSB repair tends to appear as conserved in plants, despite the fact that there can be differences that can exist. For example, the loss of a functional Mre11 complex leads to lethality in vertebrate systems while viability is maintained in plants (Daoudal-Cotterell *et al.* 2002, Puizina *et al.* 2004).

In *Arabidopsis*, the *MRE11* **gene has been observed to be required for DSB repair, but it is not required for the induction of Spo11-dependent meiotic DNA breaks** (Puizina *et al.* 2004).

The mutation of the *Arabidopsis RAD50* **gene has led to the sterility as well as hypersensitivity to the DNA-damaging agent: methylmethane sulphonate to ,in turn, suggest that a conserved role for** *AtRAD50* **in DSB repair in plant cells** (Gallego *et al.* 2001).

Considering the case of *Saccharomyces cerevisiae*, SPO11 catalyzes DSBs and this tends to initiate recombination during the process of meiosis and leads to the formation of SC. In yeast, loss of DSB leads to the loss of synapsis.

The identification of the SPO11 homologs have been done in a number of other eukaryotes, for example, *C. elegans* **and** *Drosophila.* **Considering yeast and some other eukaryotes**, it can be said that the identification of only one *SPO11* **homologue has been done; however, in** *Arabidopsis*, **there were three which have been found.**

Therefore, while initiating DSB and recombination, the SPO11 role tends to appear to be conserved in higher plants, along with *AtSPO11-1*, which further reduces the DSBs and also, the bivalent formation in prophase I in meiosis in both sexes.

Thus, the initiation of the recombination is done in *Arabidopsis* **which is done by the formation of DSBs. However, the gametogenesis is not completely prevented in the** *Atspo11* **mutant, which is as seen in** *S. cerevisiae*, despite the fact that the fertility is greatly reduced.

The identification of a conserved group of RAD51-like proteins has been done in yeast through to vertebrates. These proteins tend to play a role in recombination as well as damage repair pathways and also, they are homologous to the prokaryotic RECA protein.

The identification of the paralogues of all six vertebrate *Rad51-like* **genes has also been done in the** *Arabidopsis* **genome. The mutation of one of these,** *Atxrcc3* **leads to aberrant male as well as female gametogenesis.**

In addition to this, these plants are hypersensitive to DNA-damage treatments, and this supports the role of Rad51-like proteins in recombination repair (Bleuyard & White 2004).

There have been suggestions provided by the yeast two-hybrid assays: that RAD51 and DMC1 relate to the BRCA2 protein and furthermore, they play a role in homologous recombination and meiotic progression (Siaud *et al.* 2004).

Moreover, this is supported by the failure of meiosis as well as aberrant recombination events in RNA interference silenced *BRCA2* **lines.**

Considering the case of yeast and humans, *SKP1* **genes have been observed to regulate the mitotic cycle, but there is no role that has yet been assigned to the meiotic cycle.**

In Arabidopsis, the identification of nine *SKP1-LIKE* **genes has been done along with roles in both the vegetative as well as generative cycles. When mutated, the** *ASK1* **gene leads to male sterility and also, abnormally sized microspores because of a defect in the homolog separation** (Yang *et al.* 1999*a*).

The homologous pairing tends to occur during metaphase I; nevertheless, uneven segregation tends to occur during anaphase I along with some of the homologs which remain associated after metaphase.

The formation of spindle takes place, (as judged by immuno-fluorescence microscopy) tends to appear normal. It is important to have the yeast *SPK1* **gene for entry into the S phase and also, for the completion of the M phase of mitosis.**

The SPK1p has an important role in the ubiquitin – ligase complex, SCF which tends to target the specific proteins for ubiquitination proteolysis. It is still not known whether Spk1p has a role in the process of meiosis in either yeast or humans. According to Yang *et al.* (1999*a*), *SPKP1-LIKE1* (*ASK1*) can play a role while removing protein(s) which is required for homolog association before anaphase I.

They tend to speculate that the ubiquitin-mediated proteolysis can play a role in some other instances of chromosomal separation. Some other *SPK1 - LIKE* **genes can have particular roles related to some other developmental processes, such as vegetative growth. This has come about because of the ancestral duplication and functional divergence.**

While mutation, the *MS5/TDM/POLLENLESS3* **gene lead to 'polyads', are tetrads along with more in comparison to four pools of chromosomes, after the male meiosis because of an additional division after meiosis II** (Ross *et al.* 1997). The MS5 protein has a low level of similarity to the rat synaptonemal complex protein 1 and furthermore, has a regulatory subunit of cyclin – dependent kinase (Glover *et al.* 1998).

After meiosis fails to occur, separation of the microspores occurs in the *stud/tetraspore (std/tes)* **mutant due to a failure in male cytokinesis** (Hulskamp *et al.* 1997, Spielman *et al.* 1997).

This tends to lead to the formation of large tetra – spores that contain all four meiotic products and moreover, they go through PMI and PMII further forming a number of gametes in an enlarged pollen grain. In these, some are viable, and also, the formation of polyploidy embryos can take place through the process of fertilization by using *std/tes* **pollen.**

Considering the *std/tes* **mutant, the radial arrays of microtubule related to the male meiotic cytokinesis fail to form. The TES protein has homology to a small number of plant kinesins and moreover, it is speculated that it tends to play a role as a microtubule - associated motor. This is possibly as homo- or heterodimers which is as seen in some other kinesins** (Yang *et al.* 2003*a*).

The identification of a number of kinesin-like proteins has been done in eukaryotes. Along with this, they are thought to be involved in transport along microtubules and possibly microtubule polymerization and depolymerization.

The *TES* **is considered to be one of 61 kinesins identified in the** *Arabidopsis* **genome. This is expressed both in vegetative as well as reproductive tissues. Nevertheless, no phenotype has been in relation to vegetative development. Furthermore, it is thought that a degree of redundancy can exist in between the different kinesins.**

4.2.5. Transcriptional Regulation of Gametophyte Development

During the forming of gametes in plants with transcriptional regulators homology related to yeast and mammals, identification of several transcription factors are considered to have been done. The homology would be limited in the SPL/NZZ gene with the family of MADS box concerning transcription factors.

It would have been associated with early prerequisites of germline and for the purpose of forming patterns inside the ovule (Schiefthaler et al. 1999, Yang et al. 1999b) and would be believed to be engaged in the gene expression regulation accompanying the switch towards gametophytic development.

The homology of the MALE STERILITY1 gene would be associated with the plant homeodomain (PHD)-finger group of transcriptional regulators acting while the callose wall would be rupturing releasing the microspores.

It might be considered to be regulating the tapetal gene expression, thereby, producing pollen inside the anther (Wilson et al. 2001). In Arabidopsis, an additional PHD-finger transcription factor is supposed to have been identified, MALE MEIOCYTE DEATH1 (MMD1).

Defects related to male meiotic development would be seen in the MMD1 mutants presumably generating cell death in meiocytes (Yang et al. 2003c).

The ABORTED MICROSPORE1 gene mutation would be leading to empty anthers without any pollen resulting from the tapetum and microspores degenerating prematurely (Sorensen et al. 2003).

Homology would prevail between the AMS1 gene and the basic helix-loop-helix (bHLH) group in relation the MYC subfamily belonging to the bHLH genes and would probably control post- meiotic expression of the gene in the anther (Sorensen et al. 2003).

4.3. FEMALE GAMETOPHYTE DEVELOPMENT

The female gametophytes (mega gametophytes or embryo sacs) develop within the ovule where as the male gametophytes (the pollen grains) in the angiosperms mainly develop within the anthers.

Further basically for observing gametophyte development in a large number if diverse plant species- various development studies over a century have been conducted in order to examine all of these species and get more clarity on them (Friedman et al, 2008).

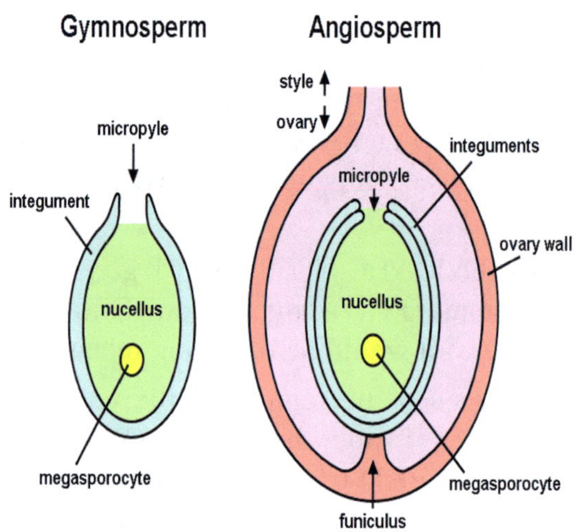

Figure 4.4: Embryo Sac diagram.

Source: Image by Flickr

A characteristic organization which is commonly termed as the Polygonum type in the embryo sac is basically seen in most of the flowering plants, including Arabidopsis and maize (Maheshwari, 1950). Moreover, it also contains seven cells of four distinct cell types. Further, there four types of cell they are described as follow:

- The cell which mainly gives rise to the embryo is the first cell out of the four and is commonly termed as the egg cell.
- The cell which important for the pollen tube attraction comes the next or simply it the second cell which is basically two accessory cells or called the synergid cells.
- The cell which gives rise to the endosperms is the third cell.
- Antipodal cells are the fourth cell, the special characteristics of these three antipodal cells is that their function is basically unknown.

Integuments are basically the diploid sporophytic tissuses and these integuments are basically used to enclose the entire embryo sac within them. Further, the other function which is performed by the integuments is that they will basically constitute the seed coat in the mature seed coat.

A Structure which is namely known as the pollen tube is formed when the pollen lands on the stigma. This pollen tube penetrates the embryo sac through an opening in the integuments called micropyle.

Further by entering the synergid cells, the micropyle basically delivers two sperm cells. And these further subsequently degenerates. Further it is observed that the other two synergid cells will be degenerated.

By these two sperm cells of the pollen tube, the two gametic cells of the embryo sac, the central cell, and the egg will definitely undergo double fertilization.

The diploid zygote is mainly obtained by the means of the fertilization of the central cell by the second sperm cell from the pollen tube and this the reason which provides the above- mentioned result.

Further, the result of the triploid endosperm is basically obtained by the process of fertilization of the central cell by the second sperm cells. Moreover, for the germinating seedlings, or as the nutritive source for the embryo, this triploid endosperm basically is functional.

The model plant Arabidopsis depicts a schematic outline of the development of a seven celled embryo sac. And this model entirely explains out to this.The archesporail cell is the one that enlarges and differentiates into the female meiocyte or megaspore mother cell (MMC) and this archesporial cell is the one in the developing ovule, a single cell from the internal tissues (which is called the nucellus).

The four spores are mainly generated by the MMC when it undergoes meiosis. Further leaving only the proximal (chalzal) megaspore as the functional megaspore (FM stage of female gametophyte stage 1) in each ovule by these above mentioned four spores.

Further designed as stages FG2 to FG5 (Christensen et al. 1997) therefore for this, the megaspores then undergoe three sequential mitotic nuclear divisions. Moreover, the differentiated embryo sac, four at each pole.

The polar nuclei will further contribute to the central cell nucleus. And this polar nuclei cell is basically formed by the means of celluarization, which furthers results in the formation of the seven cells owing to the nuclear migration of two of these nuclei.

This mainly contributes to the polar nuclei. The chalazal (proximal) end of the ovule is adjacent to three antipodal cells whereas on the other hand in seven celled embryo sac which is called the micropylar (distal) end of the ovule that is adjacent to the egg cell and two synergid cells.

A large number of vacuole and central cell occupies the embryo sac is contained by the homo diploid (2n). All the mentioned cells are the ones which basically possess the distinctive morphologies and these cells are also highly polarized within the embryo sac (Huang and Russell, 1992).

The distinction between these nuclei in their cell fate specification is basically marked by the means of the precise migration and positioning of the nuclei at the eight nucleate at FG5 (Huang and Sheridan, 1994; Webb and Gunning, 1994). The central nucleus will become one of the two polar nuclei of the central cell whereas on the other hand the remaining of the two nuclei, the more distal nucleus of the pair will be formed from the future egg cell.

The other pole of the nucleus of the central cell will be basically contributed by the means of the most centrally located nucleus of the four nuclei from the chalazal pole.

Due to the first polar nucleus as the embryo sac undergoes cellularization to form the seven celled embryo sac and also this nucleus migrates towards the micropylar pole and becomes positioned close to the first polar nucleus that undergoes the above mentioned process (Webb and Gunning, 1994).

The antipodal cells are basically degenerated at the time of fertilization whereas on the other hand, the following cellularization in Arabidopsis the polar nuclei fuse to form the homo diploidcentral cell nucleus.

Until the process of fertilization the two polar nuclei are generally unfused. Further, to form the triploid endosperm they will further fuse with a sperm nucleus (Mol et al., 1994).

4.4. MALE GAMETOPHYTE

The development of male gametophyte in higher plants would be regarded as an intricate process, requiring the synchronized involvement of different cell and types of tissues along in association with their distinct patterns of gene expression. The process would include the participation of both gametophytic (pollen) and sporophytic (anther, tapetum) tissues.

Figure 4.5: Conifer male gametophyte.

Source: Image by Wikimedia Commons

There would be two particular stages of gametophytic life cycle:

- The developmental stage occurring inside the locules resulting in forming mature pollen grains

- The functional stage also known as the programic stage, would be starting with the effect of grains on the surface of stigma and would be continuing with growth of pollen tube and would be finishing with double fertilization.

In comparison to the sporophyte, it would progressively be an easy and extremely reduced system, thus, providing exclusive opportunities for studying the developmental control of gene expression and functional interaction amongst various type of cells.

A large frame of evidence, in terms of microsporogenesis, would be showing the gametophytic gene expression as both intricate as well as widespread. Additionally, a huge segment related to such genes would be expressed in the sporophyte.

Such an act of "haplo-diploid" transcription would have vital repercussions regarding the gametophytic selection competence and its part in the successful evolution of flowering plants.

Competition pertaining to the fertilization of egg cells limited in number, among the male gametophytes, would be common and genetic variations between separate microgametophytes would be reflected through the marked differences in growth rate of the pollen tube (Mulcahy 1979, Mulcahy et al. 1996).

Gametophytic selection would be acting like an obstruction against the inadequately functioning haploid genomes and would serve to lower the effect of random events promoting severe improved haploid genotypes selection.

Such conditions claim for the rapid gametophytic gene evolution encoding dedicated functions for the improvement of male gametophytic fitness.

On the other hand, substantial overlapping among expression of genes related to both generations would be resulting in the sporophytic fitness being influenced by gametophytic selection (Mulcahy 1979, Mulcahy et al. 1996).

Recent studies regarding developing male gametophytes would categorically validate the overlapping level of the haplo-diploid more than 95% (Honys and Twell 2004, Twell et al. 2006). Thus, provisioning huge probability for the gametophytic gene selection that would be impacting the sporophytic traits.

4.4.1. Male gametophyte development in angiosperms

Angiosperms have seeds in the enclosed fruit obtained from a flower ovary. The flowers would be consisting of mainly sporophytic tissues, together with male and female gametophytes of extremely reduced size compared to all other land plants.

Angiosperms would even possess a distinct double fertilization property, generally making a triploid endosperm additionally with the embryo.

The formation of the male gametophyte would be occurring within the anthers of the stamens and the ovules inside the pistil would be the location of female gametophyte. Four pollen sacs (locules) in the anther would have several mother cells of microspore, each experiencing meiosis for the production of four microspores in a tetrad. The beginning of the male gametophyte generation would be with the microspore.

Originally, the microspore have a cytoplasm distributed systematically with a haploid nucleus being located centrally. Later, a huge vacuole would be formed at the center dislocating the nucleus towards the side.

In several flowering plants, inclusive of Arabidopsis and maize model species, two mitotic divisions would be taking place throughout the development of pollen. A huge vegetative cell and a considerably smaller generative cell would be produced by the first division.

Majority of the cytoplasm from the microspore cell would be inherited by the vegetative cell and would have a moderately loose nucleus, active in transcription and enveloping the generative cells as a whole.

However, the generative nucleus is supposed to be organized more tightly as well as, slowing down in transcription. Later, the generative cell would be experiencing a second mitosis for producing two sperm cells. Development of the pollen would be subjected to the functioning of an adjacent sporophytic tissue named tapetum.

At the time of maturation, the pollen would be dehydrated. It would be rehydrated after pollination and germination would occur producing a pollen tube. The tube grows in the direction of the ovule, delivering a passageway for the sperm cells for the purpose of reaching the female gametophyte.

Since the female gametophyte would be inside the female reproductive organ concerning the sporophyte, the pollen tube is considered to be extending to significant distances, often the pollen grain size, for reaching to the egg.

4.5. CONCLUSION

Despite the fact that the process of gamete formation in plants has many unique features, there has been much learnt from the comparative analysis in between plants and some other eukaryotic systems.

The plants tend to have a number of factors which have made them to be considered as desirable for the analysis of gamete development. These include late germline specification, the non-lethality of mutations that tend to affect the gamete development, as well as the large size of their chromosomes.

There are advantages that plants give along with the use of mutant screens in order to identify genes related to the gamete formation and these advantages have also provided access to genes which are difficult to be characterized by alternative routes.

This gave birth to a lot of novel information in relation with the processes of gamete formation in higher plants. In today's scenario, the times have now changed along with the advantages, which plants provide while serving to advanced knowledge of gamete formation in eukaryotic systems.

From the above chapter, the conclusion that can be drawn is that there are many phases and stages through which plants have to go through in order to form male as well as female gametes and furthermore to undergo gametogenesis.

REFERENCES

1. Honys, D., Reňák, D. and Twell, D., 2006. *Male Gametophyte Development And Function*. [ebook] p.1. Available at: <http://www.ueb.cas.cz/en/system/files/users/public/2006__Honys_Male_gametophyte__FOPB.pdf -1> [Accessed 27 May 2020].
2. Sundaresan, V. and Alandete-Saez, M., 2009. Pattern formation in miniature: the female gametophyte of flowering plants. *Development*, [online] 137(2), pp.179-189. Available at: <https://dev.biologists.org/content/137/2/179> [Accessed 27 May 2020].
3. Wilson, Z. and Yang, C., 2004. *Plant Gametogenesis: Conservation And Contrasts In Development*. [ebook] p.1. Available at: <http://citeseerx.ist.psu.edu/viewdoc/download?doi=10.1.1.539.8833&rep=rep1&type=pdf> [Accessed 27 May 2020].
4. Wilson, Z. and Yang, C., 2004. Plant gametogenesis: conservation and contrasts in development. *Reproduction*, [online] 128(5), pp.483-492. Available at: <https://www.researchgate.net/publication/8206962_Plant_gametogenesis_Conservation_and_contrasts_in_development> [Accessed 27 May 2020].
5. Wilson, Z. and Yang, C., 2004. Plant gametogenesis: conservation and contrasts in development. *Reproduction*, [online] 128(5), pp.483-492. Available at: <https://rep.bioscientifica.com/view/journals/rep/128/5/1280483.xml> [Accessed 27 May 2020].
6. Aouali *et al.* 2001
7. Armstrong *et al.* 2002
8. Azumi *et al.* 2002
9. Bai *et al.* 1999
10. Bhatt *et al.* 2001
11. Byzova *et al.* 1999
12. Canales et al. 2002
13. Caryl *et al.* 2003
14. Chapman 1987
15. Christensen et al. 1997
16. Couteau *et al.* 1999
17. Daoudal-Cotterell *et al.* 2002
18. Doutriaux *et al.* 1998

19. Eady et al. 1995
20. Faure 2001
21. Friedman et al, 2008
22. Gallego & White 2001
23. Gallego *et al.* 2001
24. Golubovskaya *et al.* 1993, 1997
25. Grelon *et al.* 2001
26. Heslop-Harrison 1966
27. Heyting 1996
28. Honys and Twell 2004
29. Huang and Russell, 1992
30. Huang and Sheridan, 1994
31. Hulskamp *et al.* 1997
32. Klimyuk & Jones 1997
33. Maheshwari, 1950
34. McCormick 2004
35. Mercier *et al.* 2003
36. Mol et al., 1994
37. Mulcahy et al. 1996
38. Nasmyth 2001
39. Piffanelli et al. 1998
40. Puizina *et al.* 2004
41. Ray & Langer 2002
42. Rhee & Somerville 1998
43. Ross *et al.* 1996, 1997
44. Schiefthaler *et al.* 1999
45. Schnable & Wise 1998
46. Schneitz et al. 1998
47. Scott et al. 2004
48. Siddiqi *et al.* 2000
49. Skinner et al. 2004
50. Smith & Roeder 1997
51. Sorensen et al. 2003

52. Spielman et al. 1997
53. Twell et al. 2006
54. Webb and Gunning, 1994
55. Weiss & Maluszynska 2001
56. Worrall et al. 1992
57. Wylie 1999
58. Yadegari & Drews 2004
59. Yang & Sundaresan 2000
60. Yang *et al.* 1999
61. Zhao & Garbers 2002
62. Zhao et al. 2002

Chapter 5 Pollination and Breeding

CONTENTS

The chapter named as 'Pollination and Breeding' basically focuses on the process of pollination and breeding that takes place in plants. In the first half of the chapter, it includes a general overview of pollination and also the modes of pollination.

Then it describes the agents of pollen dispersal such as insects, moths, wind, birds, water and mammals. Going further, the chapter provides insights about the history of the plant breeding and the phases that are included in the plant breeding.

In the second half of the chapter, it provides details about the methods used for plant breeding and have discussed the mating systems. In the end, it provides knowledge about the importance of the plant breeding in today's era.

5.1. INTRODUCTION

The breeding systems are considered to be important and most of the time, it is neglected aspects of reproductive biology. There is much research going on and the main research content includes floral characteristics, mating systems and pollination.

The floral biology, pollination pattern and behavior are related closely to the reproductive fitness. Therefore, these are considered to be the strong selective forces. The reproductive success of plants is limited most of the time by lack of pollination, fecundity and mate limitation and mate availability.

It is also important to have flower arrangement and reward availability which are said to be important factors. These factors tend to influence mating success.

At both the individual as well as population levels, the timing of pollen presentation and stigma receptivity can strongly correlate with mate availability.

Moreover, the species with long flowering periods can avoid pollination limitation and reproductive deficit which can be the result of an increased overlap along with insect activity.

There are individuals with hermaphrodite flowers which can resort to self – pollination. This in turn, results in maximum seed setting but can result in minimum genetic variation in their offspring. Nevertheless, geitonogamy can also mitigate pollen limitation in self – compatible plants.

Hence, the floral syndrome and breeding system investigations are conducive to understanding the pollination mechanism and mating patterns. In addition to this, they also have scientific significance for more such investigations related to the evolution and environmental compatibility of plants. Pollination is considered to be a key stage of generative propagation and the dispersal of plants is connected with the pollinators closely.

Along with this, pollination has some important application value while considering the biodiversity. Around 65 percent of flowering plants are pollinated by insects.

The insect pollination furthermore promotes the increased and rapid development of flowering plants. The pollination mediates the male – female interactions and moreover, it tends to depend on abiotic as well as biotic factors.

Considering the pollinator – dependent species, pollination success is highly influenced by the frequency of visits, the breeding systems and the intra – floral foraging behavior of pollinators. Considering the self – incompatible plants, the visitation rate and the foraging behavior of pollinators within plants tend to influence the amount of self – pollen that is being transferred. Ultimately, this can result in pollen limitation.

There can be a lack of faithful pollinators, a few pollinators or low pollination efficiency in insect-pollinated plants which can lead to low pollination intensity and moreover, decreased seed yield. In addition to this, the flower-visiting behavior, which is the result obtained, would affect the seed yield as well as the quality.

There is a direct relation in between the pollination efficiency and the pollination amount and quality. This is affected by flower - visiting behavior. Thus, the interaction among population structure, pollination efficiency and pollination behavior is considered to be very complex and furthermore, requires synthetic as well as meticulous research.

5.2. POLLINATION: A GENERAL OVERVIEW

Pollination is considered to be the process of transferring the pollen grains from the anther to the stigma. Anther is the male part of the flower and stigma is the female part. This is the process which normally precedes fertilization. It is considered to be an important process in the reproduction of plants and this is the process without which sexual reproduction will not take place.

Over the ages, pollination is considered to be a process, which has been coordinated and perfected as plants co-evolved with animals. Here, the animals tend to act as pollinators or pollination agents. The plants and animals tend to coexist in same habitats.

There exist co-evolution in between plants and animals with respect to pollination resulted in the development of pollination syndromes. There are specific pollination agents who are considered in pollination syndromes which tend to pollinate particular plants or flowers.

However, this is not a water tight arrangement because of the fact that there are polyphilic flowers that attract and moreover, are pollinated by a number of different types of pollination agents and also, there are polytrophic pollinators which are attracted as well as pollinated by various types of flowers.

Pollination is considered to be a symbiotic relationship that is possessed by the plants and the agents or animals and in this type of relationship, they both are benefitted. The agents get nutrients from the flowers and the benefit received by the plants is improved reproduction. In crop production, the latter is very important.

Angiosperms are credited with flowers that are the most beautiful gift of nature. Though scientifically of reproductive relevance, the latter have much greater significance in the ecosystem as well as for the human society.

> As Richards remarks (1987): An outstanding feature of Angiosperms is the amazing diversity in forms and colour that has been adopted by the inflorescence, sufficient to inspire great art, fuel a major industry and serve as a solace for suffering mankind. Yet, the flower is merely a sex organ, and never has any function except to promote reproduction by seed, usually sexually. The beautiful, weird, sinister, astounding forms that flowers have acquired are strictly pragmatic, and have encouraged the ecological diversification, and dominance, of the flowering plants.

This is the quote which was provided by Richards in the year 1987 and furthermore, it highlights the ecological importance of flowers. If the authentic paleontological, as well as present ethological records, can further prove (and they do) that the evolution as well as perpetuation of this floral diversity is because of the pollinators.

This, in turn surely tends to demonstrate their importance. Similarly, if in case, it is also proved that pollinators provide their help in increasing the seed yield of a number of crops, this becomes important to underline their economic importance.

Considering angiosperms, the pollination is said to be an important event which tends to act as a prerequisite for the purpose of sexual reproduction. Pollen performs the function in plants that is similar to the role of sperm that is played in animals.

Therefore, the successful pollen transfer is considered to be very important. Nevertheless, the pollen is a non-motile spore. It is important for pollen to be transferred from anthers (the destination for their production) to the stigma (the destination for their germination) by a vector.

There are a number of different plant species which tend to exercise various pollination modes and provides benefits which in turn depend upon the type of pollen being transferred.

When the self – pollination normally increases the homozygosity, it becomes important for pollen from other flowers, plants or genotypes that they increase the heterozygosity.

In general terms, it can be said to sacrifice the plant quality (specifically in outcrossers), whereas the out-crossing tends to help in order to increase the hybrid vigor. This in turn leads to healthier as well as stronger plants, even though this is not always the case. In addition to this, the type of pollination also determines the chances of gene recombination as well as exchange between individuals.

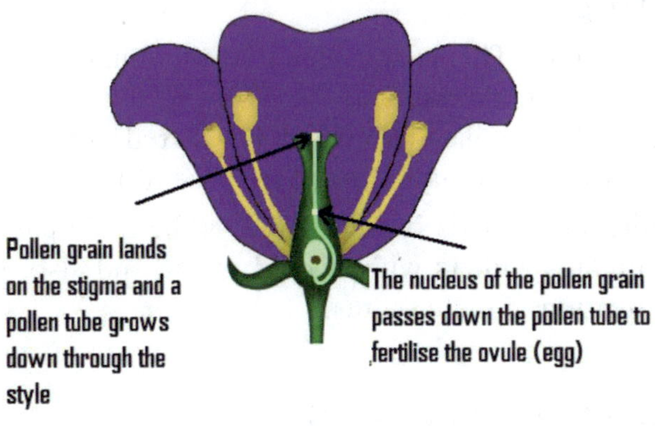

Figure 5.1: Image showing Self – Pollination.

Source: Image by Wikimedia Commons

Considering the continuously changing environments, it is important for gene recombination (and also, cross – pollination) that they provide an opportunity in order to produce strains which are better suited for new conditions. Therefore, it is considered to be an ecological necessity.

While there have been constant changes in the pest scenarios along with rapidly evolving resistance to pesticides, (especially considering the tropics), the gene recombination and heterozygosity through cross – pollinations are considered to be reliable means for increasing the crop yield. They tend to provide some specific crops with the opportunity to keep up with or escape their enemies.

5.2.1. The flower

It is important to understand the way in which the flowers work and relate to pollination as well as pollination modes, prior to exploring the benefits of pollination.

Considering the typical hermaphrodite (bisexual) flower, it can be said that it has four parts:

- Corolla - The corolla consists of flower petals. It is the colored part of the flower, and this in turn provides the primary attraction as well as stimulus for pollinators.
- Calyx: The calyx consists of sepals which is normally green. This in turn provides protection to other floral parts during the time of bud stage.
- Gynoecium: This is considered as the female part of the flower. It is the female gamete: the ovule in the ovary. The pollen is received at the distal tip which is referred to as the stigma, where the former germinates for the purpose of fertilization.
- Androecium: This is considered as the male part of the flower. It contains anthers with to hold pollen and also, function as the destination of male spores.

5.3. POLLINATION MODES

As mentioned above, pollination is said to be the process of transferring pollen from the anthers to the stigma. The agent which provokes this transfer is referred to as the pollinator. In normal scenario, angiosperms tend to exhibit two kinds of pollination mode:

- First mode is when pollination takes place within a flower and this is known as self - pollination or selfing. The process of self-pollination takes place:
- When the flowers are bisexual and the flowers tend to have stigmas and anthers at the same heights (the stamens and the style are observed to be of the same length);
- When both sexes mature at the same time (protandry or protogyny are not present); and
- When the contact of newly dehisced anthers (further releasing the fresh pollen) with the receptive stigma is imminent.

Under the above conditions, the proximity of anthers to stigma further leads to self - pollination. If in case, selfing results in fertilization, then this is important to indicate at least some occurrence of strict inbreeding.

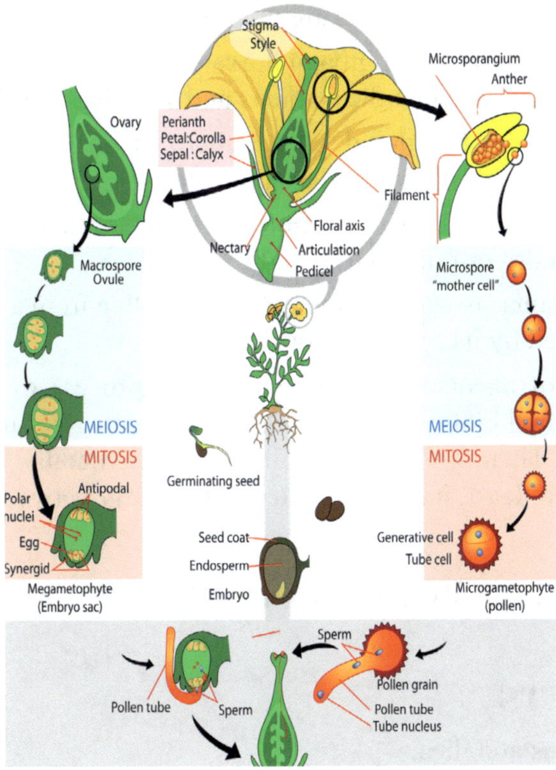

Figure 5.2: Pollination Modes.

Source: Image by Simple Wikipedia

- When transfer of pollen from one flower is done to the stigma of another through carrying, the process is referred to as cross - pollination or out - crossing. In this mode of pollination it is important to have an external agent in order to accomplish the transfer of pollen.

Mainly, there are two kinds of out - crossing:

- When the occurrence of crossing takes place in between flowers of two different plants; and
- When the occurrence of crossing takes place between flowers of the same plant (genetically, this process is equal to selfing despite the fact that, a foreign pollen vector is required).

In plants, the out - crossing is considered to be important where there can be many aspects:

- either flowers or plants are unisexual;
- anthers as well as stigma of the same flower have different heights (which means that the stamens and style have different lengths);
- during the functional phase, there is no contact of dehisced anthers with stigma of the same flower;
- sexes mature at different times (which means there is presence of protandry or protogyny); and
- the plants are self-incompatible (i.e. pollen from a plant cannot be utilized by its own flowers).

Two kinds of agents were brought by the process of out - crossing namely: abiotic and biotic. Abiotic pollinating agents refer to the inanimate physical forces. Therefore, abiotic pollination is generally "random", or it can be said that at least, it is not directed particularly between the flowers.

5.4. AGENTS OF POLLEN DISPERSAL

5.4.1. INSECTS

- Beetles and flies

Trapping insects for ensuring the pollinations was an ancient principle and this principle was re - adopted by some of the advanced families including the orchids and milkweeds. Furthermore, the elaboration perfected the flower traps of primitive families.

Let's take an example. The cuckoopint (referred to as *Arum maculatum*) tends to attract minute flies. These minute flies normally breed in cow dung and also, they use the means of a fetid smell. The generation of this smell takes place in early evening, and with this, heat is generated. This considerable heat helps in volatizing the odor ingredients.

There are a number of flies who visit the plant, and among them there are many who carry Arum pollen. Then they enter the floral trap via a zone of bristles and fall into a smooth – walled floral chamber.

From this chamber, escape is almost impossible. They gorge on a nutritious stigmatic secretion that is produced by the female flowers at the base of chamber. Thus, the flowers effect cross pollination.

At late night, the stigmas no longer function. When this happens then the male flower (which is situated at much higher position on the floral column) proceeds to bombard the flies with a rain of pollen.

The next day smell, heat and food are gone and at that time, the prisoners, who are tarred with stigmatic secretion and feathered with pollen, are allowed to escape. The escape thus takes place by a wilting of the inflorescence (that is the flower cluster).

Figure 5.3: Beetles.

Source: Image by Pixabay

Usually, the flies which escape are soon re-caught by another inflorescence that is still in the smelly as well as receptive stage and there occurs cross –

pollination again. There are superb timing mechanisms which underlie these events. The metabolic process is heat generating in the inflorescence and furthermore, is triggered by a hormone named as calorigen. This originates in the make flower buds only and also, under the right conditions. Considering the tropical plant Amorphophallus titanum, there are giant inflorescences which similarly trap large carrion beetles.

Generally, the trap flowers tend to victimize beetles or flies of a primitive type. Despite the fact that the beetles most probably were involved as pollinators, while flowering plants as a group, their performance after some time in pollination has been observed to be disappointing.

There are some modern beetles which tend to visit smelly flowers of an open type that include hawthorn and elderberry. But there are still some exceptions which are still mainly eat pollen. As a group, flies have become much more diversified in their habits in comparison to beetles.

The female short – tongued flies can be deceived by open – type flowers along with carrion smells including the examples of the flowers of Rafflesia and Stapelia. The mosquitoes have long tongues and so, are considered as effective pollinators of specific orchids (of the Habenaria species) considering the North American swamps.

Figure 5.4: Housefly cross – pollinating a flower.

Source: Image by Flickr

Considering the region of Europe, it is said that the bee fly (named as Bombylius) is an important pollinator with a long tongue. There are certain South African flies, which are extremely specialized as nectar drinkers. These flies include, *Megistorhynchus longirostris*, which has a tongue which is round 60 to 70 mm (that is equal to 2.3 to 2.7 inches) long.

The voraciousness of flower beetles shows the futility of enticing insect pollinators solely with pollen which is such an indispensable material. There are certain nectar-free flowers which use a defensive strategy that cater to beetles and bees; examples of which include wild roses, poppies and peonies in order to produce a superabundance of pollen.

There are some other plants also such as, Cassia that have two types of stamens. One of them produces special sterile pollen that acts as food for the insects. The other one from them tends to yield normal pollen in order to fertilize the ovules. Some other flowers consist of hairs or food bodies which are attractive to insects.

- Bees

Probably, bees are considered to be the most important insect pollinators in the modern world. They tend to live exclusively on nectar. They tend to feed their larvae pollen and honey (modified nectar).

They possess striking physical as well as behavioral adaptations in order to obtain their foods. These adaptations include tongues that are long as 2.5 cm (that is equal to 1 inch), hairy bodies and special pollen baskets (which are found in honeybees and bumblebees).

Karl von Frisch was an Austrian naturalist who showed that the honeybees distinguish at least four different color regions such as yellow (including orange and yellow green), blue- green, blue (including violet and purple), and ultraviolet, despite the fact that they are blind to red light.

They are very sensitive to ultraviolet rays and therefore, their sensitive enables them to follow the nectar – guide patterns which is not apparent to the human eye. They have the ability to taste a number of different sugars. Along with this, they can also be trained in order to distinguish between aromatic, sweet or even minty odors but not foul smells.

The decisive factor may be the fragrance while establishing the habit of honeybees and that is of staying with one species of flower till the time it is abundantly available. In addition to this, it is also important to know that the honeybee workers tend to establish communication between one another (both the distance as well as the direction of an abundant food source) by using special dances.

In daytime, the bee flowers tend to attract their insect visitors and this is primarily done by bright colors. At close range, this is done by special patterns and fragrances.

A number of bee flowers provide their visitors with a landing platform that is in the form of a broad lower lip on which the bee sits before pushing its way into the interior of flower. This usually contains both stamens as well as pistils.

For most of the bee flowers, hermaphroditism makes for efficiency. The reason behind this is that the flower both delivers as well as receives a lot of pollen during a single visit of the pollinator. The pollinator never travels from one flower to another without having a full load of pollen.

Only one pollination visit is allowed by the floral mechanism of a number of bee flowers. The pollen grains of most bee flowers are said to be sticky, highly sculptured or spiny while making sure their adherence to the bodies of bees.

One load of pollen contains enough pollen grains for initiating the fertilization of many ovules. Due to this, most of the individual bee flowers tend to produce many seeds.

There are flowers which depend heavily on bees and these flowers include monkshood, larkspur, Scotch broom and bleeding heart.

There are two species of bees namely bees (Nomia) and leaf-cutter bees (Megachile) which are both efficient pollinators of alfalfa as well as they have no fear while triggering the explosive mechanism (not like honeybees). This in turn, liberates a cloud of pollen in alfalfa flowers.

In some of the Ecuadorian orchids (Oncidium), the pollination is done by make bees of the genus Centris. The beelike flowers are attacked headlong while vibrating the breeze by the strongly territorial males. They mistake them for competitors.

There are other South American orchids that are nectarless but they are very fragrant. These are visited by male bees (Euglossa species) and the reason for their visit is still not understood. The male bees (Euglossa species) tend to collect an odor substance from the surface of the flowers which is then stored by them in the inflated parts of their hindlegs.

- Wasps

There are a few wasps that tend to feed their young pollen or nectar. However, the yellow jackets occasionally occur in large numbers and moreover, they tend to visit the flowers in search of nectar for their own consumption.

This can be assumed local importance as pollinators. The preference of these insects is the brownish - purple flowers which have nectar that is easily accessible, such as those of figwort.

Figure 5.5: Wasps.

Source: Image by Needpix

Some Mediterranean and Australian orchids tend to mimic the females of specific wasps (of the families Scoliidae and Ichneumonidae). This mimicking is done so successfully that the males of these species tend to attempt copulation and moreover, that receive the pollen masses on their bodies.

Considering figs, it can be said that it is not the sexual drive of pollinator that is harnessed by the plant. But it is the instinct to take care of the young as well as tiny gall wasps (Blastophaga) use the diminutive flowers (within their fleshy receptacles) which act as incubators.

5.4.2. Butterflies and moths

Only by the development of the modern flower which provides their food, the evolution of moths and butterflies (that is Lepidoptera) was made possible. Nearly all species of Lepidoptera consist of a tongue, or proboscis, which is especially adapted for the purpose of sucking.

The proboscis is coiled at rest as well as it is extended while feeding. While they feed, the hawkmoths hover; on the other hand, butterflies alight on the flower.

Figure 5.6: Butterfly pollinating on flower.

Source: Image by Wallpaper flare

Significantly, there are some butterflies who can taste sugar solutions by using their feet. Despite the fact that moths are nocturnal and butterflies are diurnal (in general sense), there has been demonstration of a color sense while representing both of them.

Generally, there is similarity in the color sense in Lepidoptera and the sense in the bees. But there are swallowtails and some other butterflies which tend to respond to red colors.

There are typical moth flowers such as stephanotis, honeysuckle and jimsonweed. These are light -colored and are most of the time long and narrow without landing platforms. Sometimes, the petals are fringed; most of the time the copious nectar is in a spur. At night, they are open as well as overwhelmingly fragrant.

Typically, color and fragrance tend to provide cooperation while guiding Lepidoptera to flowers. But there are some cases where strong emphasis has been put on just one attractant. Let's take an example. There are certain hawk moths that can find fragrant honeysuckles which are hidden from sight.

The butterfly flowers include those of butterfly bush, verbena and milkweed. They are said to be conspicuously colored and most of the time they are red. In general, they are smaller in comparison to moth flowers. But when they are grouped together in erect as well as flat – topped inflorescences then they further provide landing space for the butterflies.

Figure 5.7: Verbena Flowers.

Source: Image by Pixabay

There are some important pollinating moths which are the various species of the genus Plusia. Sometimes, they are said to be occurring in a large numbers. In daylight, the humming bird hawkmoth (Macroglossa) are much active. There is a small moth named *Tegeticula maculate* which tends to present an interesting case.

This entirely dependents on yucca flowers; in their ovules, the development of larvae takes place. Prior to depositing their eggs, the flowers are pollinated by the females while further following a pattern of specialized behavior which is almost unbelievable. This includes preparing a ball of pollen grains and then, carrying it to the stigma of the plant that they are about to use for the purpose of egg- laying.

5.4.3. Wind

Despite the fact that the wind pollination is prevalent in primitive cycads and in conifers like pine and fir, in the flowering plants, the wind pollination is considered to be a secondary development. Most probably, it arose when such kind of plants left the tropical rain forest where they originated as well as faced a more hostile environment.

In this, the wind weakened the effectiveness of smell which acts as an insect attractant and the lack of pollinating flies and beetles made itself felt too.

Wind pollination is considered to be a wasteful process due to lack in precision. Let's take an example. One male plant of *Mercurialis annua*, which is a common weed, tends to produce 1.25 billion grains of pollens that is to be dispersed by the wind.

A male sorrel plant tends to produce 400 million. Generally, despite the fact that the concentration of such pollen becomes very low about 0.4 km ($1/4^{th}$ mile) from its source, however, in windy areas it can cover considerable distances.

An example can be of pine pollen. This is naturally equipped with air sacs and this can further travel up to 800 km (500 miles), despite the fact that the grains can lose their viability in the process. Still, this gives only a slim chance in a statistical manner that an individual stigma will be hit by one or more than one pollen grains. In addition to this, there is a fact that is relevant to the number of pollen grains per stigma and that is the dry, glueless as well as smooth – surfaced grains are shed singly.

Since, there is reduction in the number of fertilizing pollen grains; there occurs reduction in the number of ovules in a single flower. This results in the number of seeds in each fruit. For example, in hazel, beech, walnut and oak, there are only two ovules per flower. In stinging nettle, birch, elm, sweet gale and grasses, there is only one ovule per flower.

The flowers that are wind – pollinated are inconspicuous which is being devoid of insect attractants rewards which include fragrance, nectar and showy petals. Most of the time, the blooming takes place before the leaves are out in spring or the flowers can be placed at high on the plant in order to facilitate exposure of the flowers to the wind.

Flowers, inflorescences or the stamens move easily in the breeze. This tends to shake the pollen or the pollen containers (anthers) that burst open in an explosive fashion when are hit by the sun, further scattering the pollen widely into the air. Most of the time, the stigmas are long and divided into arms or lobes, so that a large area is required in order to catch pollen grains.

Furthermore, considering the open areas, it can be said that the wind - pollinated plants of one species grow together most of the time in dense populations. There are high chances of self – pollination and these chances are minimized by the fact that a number of species are dioecious or hazel-like and have separate male as well as female flowers on each plant.

There are many flowering plants that depend on wind pollination and some familiar ones from them are sedges, grasses, rushes, cattail, lamb's-quarters, hemp, sorrel, nettle, alder, hazel, plantain, birch, poplar and oak. However, the tropical oaks may be insect-pollinated.

5.4.4. Birds

In Europe, the beginning of the study of mechanisms of pollination takes place, where the pollinating birds are rare. Because of this, their importance is underestimated most of the time. In fact, considering the tropics and the southern temperate zones, the birds are as important as pollinators as insects are or even more so.

There are around a third of 300 families of flowering plants which have at least some members along with ornithophilous (bird – loving) flowers that is the flowers which are attractive to birds. In contrast, around 2000 species of birds belong to 50 or more families. They tend to visit flowers less or more regularly in order to feed on nectar, pollen and flower – inhabiting insects or spiders.

There have been special adaptations to this way of life and that too in the form of slender (and sometimes curved). The beaks and tongues along with brushes or shaped into tubes are found in more than 1600 species of eight families.

These eight families include hummingbirds, honeyeaters, sunbirds, brush-tongued parrots, white – eyes, honeycreepers (or sugarbirds), flower – peckers and Hawaiian honeycreepers like the iiwi.

Figure 5.8: Hummingbird pollinating red honeysuckle flower.

Source: Image by Pikrepo

In general terms, the sense of smell in birds is poorly developed. This sense is not used in their quest for food. Rather than this, they tend to depend on their powerful vision as well as their color sense which in turn resembles that of human.

Moreover, the sensitivity of eye of the bird is considered to be the greatest in the middle and red part of spectrum. Sometimes, this is ascribed to the presence in the retina of drops of oil which is orange – red in color. This together can act as a light filter.

Despite the fact that there are explanations which have been forwarded, usually, the special red sensitivity of the bird eye is considered to be the reason why a number of bird – pollinated flowers are of a uniform as well as pure red in color. The combinations of complementary colors including orange and blue or green and red are found as are white flowers.

As per the expectation, the bird flowers generally tend to lack smell. In the daytime, they are open and are bigger in comparison to most insect flowers. They have a wider floral tube. Also, the bird flowers are sturdily constructed as a protection against the probing bill of the visitors.

Along with this, the ovules are kept out of the harmful way in an inferior ovary beneath the floral chamber or are placed at the end of a special stalk or behind a screen which is further formed by the fused bases of the stamens.

The latter is so strong most of the time so as to resemble the metal wire. Usually, they are in a large number, brightly colored and protruding. This is so that they touch a visiting bird on the breast or head as it feeds.

Most of the time, the pollen grains stick together in clumps or chains along with the results that a single visit can result in the fertilization of hundreds of ovules.

Hummingbirds usually suck the nectar of flowers on the wing in the Americas. Often, ornithophilous flowers (such as fuchsias) are pendant as well as radially symmetrical furthermore, lacking the landing platform of the typical bee flower.

Considering the cases of Africa and Asia, the bird-pollinated flowers are often erect and also, tend to offer their visitors nectar. They do not hover, either a landing platform or special perches in the form of small twigs near the flower.

The pollinating birds are considered bigger in comparison to insects. They have a very high rate of metabolism. Despite the fact that some hummingbirds go into a state that resembles hibernation every night while curtailing their metabolism drastically, others keep late hours. Therefore, generally, the birds require much more nectar per individual while comparing it to insects.

The bird-pollinated flowers tend to produce nectar copiously. For example, the nectar is a thimbleful in each flower of the coral tree and a glassful in flowers of the spear lily (Doryanthus). There are plants bearing typical bird flowers which are cardinal flower, red columbine, fuchsia, hibiscus, strelitzia, trumpet vine and eucalyptus. There are many that are members of the pea, cactus, orchid and pineapple families.

5.4.5. Mammals

Generally it is said that Madagascar is the country in the East Africa, in which the mouse lemurs (Microcebus), are only ten centimeters (four inches) long, and they use to obtain their food from flowers and on the other hand in Australia the diminutive marsupial honey possums and pygmy possums also are flower specialists.

There are two highly specialized tropical bats, particularly Macroglossus and Glossophaga, which also obtain most or all of their food from flowers. Small bats with sharp snouts and long describes the macroglossus (big-tongued) species of southern Asia and the Pacific, which have extensible tongues, and for picking up a sticky mixture of nectar and pollen they use to carry special projections (papillae) and sometimes a brush -like tip.

Considerably, they are considered as almost toothless. It has been observed that color sense and that sonar sense so well-known in other bats, which seem to be lacking.

They feed at night even though their eyesight is not keen but, by their highly developed sense of smell they are probably guided to the flowers principally.

With the help of their thumb claws bats hook themselves into the petals and stick their slender heads into the flowers, extracting protein-rich pollen with their tongues and viscid nectar. So, in the process of evolution the plants which are involved, responded to the bats by producing large (sometimes huge) amounts of these foods.

For example, there is one balsa-tree flower, which may contain a full 10 grams (0.3 ounce) of nectar, and one flower from a baobab tree has about 2,000 pollen-producing stamens. Succulent petals or special food bodies to their visitors is also provided by the some bat-pollinatedflowers.

It has been seen that there are some flowers that are often placed on the main trunk or the big limbs of a tree (cauliflory); or, borne on thin, ropelike branches, as it has been considered as an another striking adaptation, as they dangle beneath the crown (flagelliflory).

The same purpose is served by the pagoda shape of the kapok tree that is the facilitation of the bat's approach. Drab color, large size, sturdiness, bell-shape with wide mouth and, frequently, a powerful rancid or urine -like smell are some of the characteristics of the flowers, as they themselves include it.

The giant saguaro cactus and the century plant (Agave) are the one which are pollinated by bats, although not exclusively, and cup-and-saucer vine (*Cobaea scandens*) is the direct descendant of a bat-pollinated American plant. Calabash, candle tree, and areca palm also have bat-pollinated flowers.

5.4.6. Water

Although, pollen grains can be made to germinate in aqueous sugar solutions, it has been seen that water alone in most cases, has a disastrous effect on them. So, accordingly, only a very few terrestrial plants, such as the bog asphodel of the Faroes, use rainwater as a means of pollen transport.

Even in aquatic plants, the water is said to be seldom, as it is consider as the true medium of pollen dispersal. Thus, plants that grow only on rocks in rushing water, flower in the dry season when the plants are exposed; pollination occurs with the aid of wind or insects or by selfing are the famous Podostemonaceae.

Sending of its male and female flowers to the surface separately is done by the aquatic plant, ribbon weed. Thus, into minute sailboats the former transform themselves, until they collide with the female flowersthat are driven by the wind.

The pollen itself is dispersed on the water's surface in the Canadian waterweed, and also in pondweed (Potamogeton) and ditch grass (Ruppia), as it is however, still water-repellent. The pollen grains are wet by water in the true water dispersal (hydrophily), in which, is found only in the hornworts and eelgrasses.

5.5. PLANT BREEDING

The production of new, improved crop varieties for use by farmers is plant breeding, or crop genetic improvement. Higher yield, improved grain quality, increased disease resistance, or be less prone to lodging are some features that the new variety can have. Ideally, it is said that there will be a new combination of attributes which are significantly better than the varieties which are already available.

Moreover, the new variety will be defined as a new combination of genes which the plant breeder has put together from those available in the gene pool of that species. Genes already existing in other varieties of the same crop it may also contain, or from other distant plant relatives it may contain genes, or genes from unrelated organisms inserted by biotechnological means.

Figure 5.9: Plant Breeding.

Source: Image by Flickr

To produce the new variety, the breeder will have employed a range of techniques. After the breeder first creates the new gene combination,, elimination of thousands of poor performers is done.

There are some of the more important genetic principles that define how plant breeding occurs and the techniques breeders use, as this chapter is concerned with describing this only. It is said that plant breeding is time-consuming and costly. To proceed from the initial breeding stages to commercial release it typically takes more than ten years for a variety.

In most of the field crops where considerable genetic improvement has already occurred (e.g. wheat, barley, maize, cotton)- most new varieties exhibit an improvement of 5% or less over the nearest commercial rival. New plant varieties are most commonly encountered in the major crop species in Australian agriculture.

However, for all other horticultural crops the same principles are applied, tree and pasture species. To reach maturity before the value of the new gene combination can be assessed in perennial species the individual plants may take many years. And, this will provide greater challenges for the breeder and this only serves to lengthen the breeding program.

To certain growing areas and conditions, the advantage of a new plant variety may be specific, or it may have attributes that are not required in other regions. Among many other characters bread wheat varieties for South Australia require, boron tolerance and resistance to Cereal Cyst Nematode (CCN) because many soils in the region have high boron levels, and CCN populations can cause significant yield losses.

The edaphic and biotic stresses are different in New South Wales, as there are new wheat varieties which must have acid-soil tolerance and resistance to the fungal disease *Septoria triticii* blotch.

For any new variety the list of requirements increases, as to have a reasonable chance of isolating a new, improved gene combination (genotype) the breeding program must handle more and more material. Depending on the species involved and the aims of the program a given level of resources (labor, land, operating funds) and program will have an upper size limit.

Consequently, for a crop in a single country there are often several breeding programs. For the market share the programs may compete head-to-head, or may target different niche markets, especially with some commercial interest in that market and if they are funded by an organization.

With plant pathologists, entomologists, biochemists, agronomists, seed production professionals, molecular biologists, statisticians, and computer scientists a breeder will typically collaborate closely.

A breeding program will use the latest proven technology and an efficient and productive breeding program will draw on these disciplines. Breeding programs are highly mechanized and employ the latest bulk-handling techniques in the developed world.

At present, the population of the world is increasing at an alarming rate. On the other hand, the land area that is available for farming is gradually decreasing and the key reason behind this is urbanization, increase in salinity, soil erosion and acidity. Over the past five decades, the terms of trade in agricultural commodities have been declining steadily.

Pests and diseases of the crop plants are rapidly evolving to overcome the resistance that exists in the current varieties. Therefore, the process of plant breeding is a race to construct or synthesize new genotypes in order to maintain or increase the production per unit area of land with decreased inputs and increased quality of the product.

From the perspective of an economy, the adaptation of a new plant variety is regarded as cost-neutral to the farmer apart from seed costs. The rigorous economic analysis has depicted that public investment in the process of plant breeding pays much high returns and hence, all of the successful crop industries are under-pinned by effectively resourced breeding programs.

In order to save money and deliver significant varieties, breeders these days are making effective use of a range of techniques to speed up the process. On the other hand of the world, off-season nurseries may be used to grow two generations per year (if, in case, there are no quarantine restrictions).

The early generations of the program may be grown in the glasshouse or growth-room to attain pure-breeding genotypes as early as possible. Any novel variety is always going to be a key part of farming system and will only attain its genetic potential if the process of agronomy along with some other farming technology is in place.

5.6. HISTORY OF PLANT BREEDING

Plant breeding is an ancient activity. It dates to the very beginnings of the activity of agriculture. Just after the earliest domestication of the cereal grains, humans started recognizing the level of excellence amongst the plants in their fields and saved seed from the best for the plantation of new crops. Such kind of tentative selective methods or techniques were the signs of early plant-breeding procedures.

The outcome of early plant-breeding processes was conspicuous. Most of the varieties at present are modified from their wild progenitors that they are not capable of surviving in nature. In certain cases, the cultivated forms are quite different from the current wild relatives making it highly difficult to identify their ancestors.

Such remarkable transformations were attained by early plant breeders in a very short period of time from an evolutionary perspective, and the rate of change was much greater than for any other evolutionary event. In the mid-1800s, Gregor Mendel drew the principles of heredity by making use of pea plants and provided the important framework for scientific plant breeding.

In the early 20th century, the laws of genetic inheritance were further delineated. Due to this, a beginning was made towards applying them to the plant's improvement. One of the key facts that evolved during the short history of scientific breeding is that a huge wealth of genetic variability exists in the plants of the world and that only a start has been made in tapping its potential.

5.7. PHASES OF PLANT BREEDING

The process of plant breeding comprises four major phases, all of which occur concurrently in an established program:

Phase 1: at first, the breeder recognizes the needs or requirements of the farmers along with the deficiencies in the current varieties. However, new or improved disease resistance is needed or increased size of the seed, or just increased yield in order to make the crop profitable.

Then the breeder will collect together different genotypes that have the attributes which is highly required. This may further need screening of the germplasm collections that are available or obtaining from other breeder's genotypes.

If the required attribute is not available in the gene pool of species then the breeder may use gene technology so as to obtain the gene from anywhere else, or as a last option, breeders consider synthesizing a completely new gene. Such options will be long-term strategic decisions based on the value of overcoming the facilities and the deficiency.

Phase 2: the second phase of the breeding program is to hybridize the identified parents to bring the genes for the required attributes together in the same hybrid individual. The particular procedures will largely depend on the species that are involved, and any pre-existing knowledge of the genetic control of the attributes.

Phase 3: In the early segregating generations, the breeder selects the progeny of the crosses so as to remove those with inferior or undesirable genotypes, gradually moving towards a smaller number of elite lines. This third phase is the largest part of a breeding program. This phase involves the identification of the products of genetic segregation as well as recombination, and finding out the best of the bunch as early as possible, at the same time reducing the risk of failing to retain a superior line. Different selection processes are adopted by the breeders.

Phase 4: the final phase of breeding mainly comprises the worth of new genotype over the current varieties, bulking up enough quality seed for distribution to the farmers and eventually release of the novel variety.

The last phase consumes critical breeding resources since, though only a small number of advanced lines remain in the program every year, they have to be assessed in an extensive field trial program at various locations, and large seed quantities produced.

Breeders regularly face the dilemma of having different varieties released as early as possible while still having reliable data on regional performance of specific varieties and probable performance over a range of years.

Breeders are aware that a farmer requires convincing data so as to change the variety and feel a potential responsibility to provide information that is as accurate as well as complete.

5.8. METHODS OF PLANT BREEDING

5.8.1. Mating systems

Angiosperm mating systems decentralize around the kind of pollination, or transfer of pollen from one flower to another flower. A flower is self-pollinated if the pollen is transferred to the flower from any other flower of the same plant and cross-pollinated if the pollen comes from a flower on a different plant.

Almost half of the more effective cultivated plants are naturally cross-pollinated, and their reproductive systems consist of different devices that motivate cross-pollination- for example, protandry (pollen shed before the ovules are mature, as in the carrot and walnut), dioecy (male and female parts are borne on different plants, as in the date palm, asparagus, and hops), and genetically determined self-incompatibility (inability of pollen to grow on the stigma of the same plant, as in white clover, cabbage, and many other species).

Some other plant species, such as a high proportion of the critical cultivated plants such as barley, wheat, peas, rice, tomatoes and beans, are really self-pollinating. There are quite few reproductive mechanisms that encourage self-pollination; the most positive of which is the failure of the flowers to open (cleistogamy), as in certain violets.

In wheat, barley and lettuce the pollen is shed before or at the time when flowers open. While in the tomato pollination follows opening of the flower, however the stamens form a cone around the stigma. In such species, there is always a risk of unwanted cross-pollination.

In the controlled procedures of breeding, it is important that the pollen from the desired male parent, and no other pollen, reaches out to the stigma of female parent. When pistils and stamens occur in the same flower, the anthers should be removed from the flowers selected as females before shedding of pollen.

Figure 5.10: An image showing Pistils and Stamens in a flower.

Source: Image by torange.biz

In general, this is done with scissors or forceps. Protection should be provided from the foreign pollen. The common method which is used is to cover the flower with a paper or plastic bag. When the stigma of the female parent becomes receptive, pollen from the desired male parent is transported to it, usually by breaking an anther over the stigma. Then, the protective bag is replaced.

The production of few hybrids is hence tedious as well as expensive as it generally requires a series of delicate, exacting, and properly timed hand operations. When male as well as female parts occur in different flowers, as in corn (maize), controlled breeding is much easier.

A cross-pollinated plant having two parents, each of which is likely to vary in many genes, produces a variety of population of plants hybrid (heterozygous) for many traits.

A self-pollinated plant, which has only one parent, produces a more uniform population of plants pure breeding (homozygous) for many traits. As a result, in contrast to out breeders, there are more chances that self-breeders become homozygous and hence true breeding for a specified trait.

5.9. THE BENEFITS OF PLANT BREEDING

In the middle of the organic food boom, as well as romantic glorification of nature, in the today's scenario, plant breeding is in a difficult position. Most of the time, one forgets that humans owe their very survival to the supply of food. In the today's world, there is almost nothing, which was originally found in the natural world.

Humans can survive without a number of things but they cannot survive without food. Eventually, each and every food consumed by a human (so as to get the biochemical energy) is produced by plants. In the world, every plant can perform photosynthesis but humans cannot.

Even if there are many advanced plant breeding methods, this cannot transform a daisy into food. It should not be done either because the ancestors of human have discovered many ways a long time ago so as to use other plant species which are naturally better suited to this.

5.9.1. The Roots of Human Culture

Earlier than around 10,000 years ago, humans were considered to be the best individual specimens from the precursors of maize, wheat, rice and apples in today's time. Humans have tried to make improvements or even make them more resistant.

This is done by forcing them to breed along with the either same or different varieties. The human selection was the law of the land for many years. Its success has resulted in greater yields as well as better tasting food.

> Later in the period, the Bible vowed 'subdue the earth'. Around 2000 years ago, the apple tree had become a cultivated chimerism already where the grafting of the fruit – bearing stem is done on the root system of a different variety and this is a highly advanced biotechnological process for that time.

5.9.2. Cultivated Plants for Human Consumption

In a gradual manner, the humans cultivated some of the modern crops from their wild progenitors. In today's era, the wheat and spelt, which is consumed by human, are hybrids and furthermore, it is created from crossing the genomes of three wild species.

The maize with its oversized cobs is considered to be a descendent of the shrubby teosinte. In addition to this, the clones of apples, vines and other fruit trees experienced a loss of innocence; there are no longer forest and no meadow flora. Since, they were grafted on to rootstock; they have become capable of keeping soil-borne pathogens in check.

Figure 5.11: An image showing Maize.

Source: Image by Pixabay

5.9.3. Royalty under Threat

Such types of optimized plants were considered as a blessing at that time. This was because of the steadily increasing yields, due to which the ancestors of humans were able to do advancement both economically as well as culturally. The population of the world grew, and the "big three" which include wheat, rice and corn, continue to provide more than half of our food in the today's time. These were named as 'calorie kings' who cannot be replaced. Humans tend to depend on them, and now, this has become a curse.

The crops tend to have a serious problem and so humans too do. They are constantly required to fight off disease. Despite the fact that the same could be said for all living things, it is specifically acute for the agricultural all-star athletes. The reason behind this is that the more humans grow them, the more aggressive pests as well as pathogens become and therefore, increasing the risk of disease.

5.9.4. Creating Resistance

In today's scenario, the plant breeding is considered to be one of the most important weapons. It is something that is not than a highly selective dating service. This creates new varieties that are resistant towards disease done through selection over many years but this requires great efforts too.

For a number of years, scientists have been putting their efforts in order to make crops permanently resistant but unfortunately, there have been no result of that yet. However, an avoidance of resistance improvement is not an option. Stagnation can result in regression which can further create a future without breeding similar to a world without medicine.

This can also be the reason of the high hopes of researchers for modern breeding methods that include Crispr. These are the genome editing processes that further allow the individual genes to be particularly modified or deactivated without using foreign genetic material.

This ,in turn, enabled the development of highly resistant plants which are genetically indistinguishable from varieties which are conventionally bred. There can be organic farmers who can be even interested in such type of development. In the current scenario, they are required to treat the potatoes along with large amounts of copper so as to keep late blight in check. It would be a blessing to have a resistant variety, even if it came about by using the genome editing.

5.9.5. The Right Path to More Sustainability

In the world today, there is almost nothing which is whether organic or conventional and are found in nature around 10,000 years ago. Today's major crop varieties are required to be adapted in order to survive in the continuously changing environment.

In addition to this, fields will require to become diversified. It is important to have more plant species in order to feed the world and therefore, it is important to have more resistant varieties of such species.

Furthermore, the varieties are required to be of top -quality plus yield and conserve soil and water. Thus, it also require as little fertilizers and pesticides as possible. This will create an ideal path in the direction towards sustainable agriculture and this will also need both new plant breeding methods as well as ecological cultivation of the crops.

5.10. CONCLUSION

The pollinators are required for the purpose of ecosystem services as well as crop production productivity. Due to the global warming and the agricultural production systems, there can be changes in ecosystems; this will require to be studied and managed so as to keep the ecosystem productivity and crop production sustainable and moreover, to feed the world population that is increasing.

The conclusion that can be drawn from the above chapter is that for producing the plants which are more useful for humans requires application of genetic principles and plant breeding. These are the processes that tend to be repeated after a number of generations to change the hereditary makeup and value of a plant population far beyond the natural limits of previously existing populations.

REFERENCES

1. 2018. The Pollination Of Cultivated Plants Extension Of Knowledge Base Pollination Services For Sustainable Agriculture A Compendium For Practitioners. [ebook] p.2. Available at: <http://www.fao.org/3/i9201en/I9201EN.pdf> [Accessed 12 May 2020].
2. Allard, R., 1999. Plant Breeding | History, Applications, & Methods. [online] Encyclopedia Britannica. Available at: <https://www.britannica.com/science/plant-breeding> [Accessed 12 May 2020].
3. J.D. Meeuse, B., 1999. Pollination - Agents Of Pollen Dispersal. [online] Encyclopedia Britannica. Available at: <https://www.britannica.com/science/pollination/Agents-of-pollen-dispersal> [Accessed 12 May 2020].
4. Luckett, D. and Halloran, G., n.d. PLANT BREEDING. [ebook] p.3. Available at: <https://cdn.csu.edu.au/__data/assets/pdf_file/0007/2805559/Chapter4_LuckettHalloran.pdf> [Accessed 12 May 2020].
5. Mangena, P. and Mokwala, P., 2018. Introductory Chapter: Pollination. Pollination in Plants, [online] Available at: <https://www.intechopen.com/books/pollination-in-plants/introductory-chapter-pollination> [Accessed 12 May 2020].
6. Walter, P., 2016. The Benefits Of Plant Breeding. [online] Ethz.ch. Available at: <https://ethz.ch/en/news-and-events/eth-news/news/2016/12/the-benefits-of-plant-breeding.html> [Accessed 12 May 2020].
7. Yang, H., Zhang, R., Song, P. and Zhou, Z., 2017. The Floral Biology, Breeding System And Pollination Efficiency Of Schima Superba Gardn. Et Champ. (Theaceae). [ebook] Fuyang, p.1. Available at: <http://www.mdpi.com/1999-4907/8/10/404/pdf-vor> [Accessed 12 May 2020].

Chapter

6

Endosperm and Embryology

CONTENTS

This chapter named 'Endosperm and Embryology' focuses on the tissue named as endosperm that surrounds the embryo and helps in nourishing that. In the first half of the chapter, it describes the tissue endosperm including its origin, function, types and structural diversity. Next, it provides insights of the development of the endosperm, which includes different stages. In the second half of the chapter, the readers will get to know the relationship possessed by the endosperm, embryo and maternal tissue and also about the future aspects in endosperm research.

6.1. INTRODUCTION

There is the ability that is found in seeds that is to sense ,as well as, respond to the environmental factors which include temperature (Yamauchi et al. 2004, Toh et al. 2012), light (Oh et al. 2007), water (Preston et al. 2009, Bai et al. 2012) and nutrients (Matakiadis et al. 2009).

This is for controlling the precise timing of germination. Once there is an extension in the mechanical constraint of the tissues that are surrounding which include endosperm due to the embryonic growth potential, the completion of germination (Finch-Savage and Leubner-Metzger, 2006).

There was a study done by Finch-Savage in the year 2007. It was reported that different germination stimuli (nitrate; light and temperature, after ripening) tend to trigger some similar changes in the transcriptome.

There were suggestions provided that such kind of environmental factors further regulate some common downstream events, possibly targeting plant hormone metabolism and signaling.

Supporting this view, it has been shown that there can be regulation due to environmental signals of the hormone metabolism in the seed, and also, the seed responsiveness to hormones (Seo et al. 2009, Nambara et al. 2010, Kim et al. 2013).

In a number of plant species, seed germination is regulated by two antagonistic plant hormones named ABA and gibberellins. The involvement of ABA is there in the induction and maintenance of seed dormancy and furthermore, inhibits the seed germination (Nambara et al. 2010).

During the process of seed development, the production of maternal as well as embryonic ABA is said to be important in order to prevent precocious germination (Karssen et al. 1983, Kanno et al. 2010).

Seed germination is promoted by gibberellins (Yamaguchi 2008). Moreover, both gibberellin metabolism and signaling are influenced due to the developmental as well as environmental factors (Ogawa et al. 2003, Rodríguez et al. 2012).

There are some other hormones and regulators (that include reactive oxygen species and small regulatory RNAs) which further provide contribution to the fine – turning of seed dormancy and germination (Nonogaki, 2010, Fernández-Arbaizar et al. 2012, Leymarie et al. 2012, Toh et al. 2012).

Along with the endogenous signals, the regulation of the seed germination is done by external signals like allelochemicals. These are secreted signals that further influence the growth, as well as development, of surrounding organisms. The phytotoxins which are derived from plant can be secreted and moreover, act as allelochemicals.

Let's take an example. The sweet gale fruit leachate consists of myrigalone A (MyA), which I a C-methylated dihydrochalcone. Oracz, in the year 2012 demonstrated that MyA tends to modulate endosperm weakening of surrounding cress seeds further by regulating gibberellin biosynthesis, using cress (that is *Lepidium sativum*) seeds as a target.

There has been increased knowledge of endosperm function which has made it to be considered to be an attractive target for the purposes of breeding and biotechnology.

Specifically, the endosperm can be modified or even, manipulated in order to accumulate a number of different metabolites or proteins, because of the fact that the endosperm in cereals tends to act as the primary storage organ for the seed (Paine et al. 2005, Li et al. 2013, Wang et al. 2013, Tuncel et al. 2014).

Along with this, the endosperm tends to express unique genes that have specialized functions in stress tolerance. These kind of endospermic proteins have been used in order to acquire stress tolerance while expressing a heterologous organism (Amara et al. 2012).

In cereals, it has been well documented concerning the role of the endosperm in seed germination. The scutellum of embryo tends to synthesize as well as secrete gibberellins to the aleurone layer of the endosperm.

The induction of gibberellins takes place while synthesizing α-amylase in the aleurone layer. This in turn, secretes such hydrolysis enzymes to the starchy endosperm.

The sugars, which are released by starch degradation, are used by the embryo for its growth. The embryo – less half seed is the dissected aleurone layer as well as starchy endosperm from the embryo. This has traditionally been used in order to examine the function of the endosperm in cereal physiology as well as biochemistry.

Considering the recent time, the dissection of endosperm is being used in order to study the endosperm in some other species (Muller et al. 2006, Penfield et al. 2006, Lee et al. 2010).

In seed biology, at the moment, a major focus is put on the elucidation of the function of individual cell types or tissues. In addition to this, the focus is also on the mechanism using which these tend to interact to coordinate a systemic response.

Keeping this in mind, it is important to identify the sites (that is cell types, tissues or organs) of environmental sensing, signal synthesis and perception in order to understand the molecular mechanisms of seed germination.

There have been current advancements in crop functional genomics that have produced novel tools, methodologies as well as various new concepts. These have further allowed for comparing the conserved and divergent functions of each and every organ.

Moreover, this has promoted the accelerated translation of the concept of seed biology to be used in agriculture as well as biotechnology (Martinez-Andujar et al. 2012, Nambara and Nonogaki 2012).

During the life cycle of a plant, there is one of the most critical events that is survival during germination. There are a number of types of structural as well as functional specializations that occur in seeds. This tends to increase the probability of survival.

Specialization is the storage of seed reserves is considered to be very important. The endosperm is said to be a seed storage tissue which is formed within the embryo sac. This is formed from a second fertilization within the central cell (Esau, 1977).

Three nuclei (usually haploid) are involved in the fertilization event. In these, two are from the central cell and one is from the sperm cell. The endosperm is said to be a tissue which is unique to flowering plants. Despite this, it is not always present in all the angiosperm seeds. In flowering plants, along with other characters, the variability in seed structure, as well as development, is considered to be enormous.

The seeds of a number of species have endosperm at the early stages of seed development. The seeds do not contain endosperm at seed maturity because of the fact that it is used during the process of seed development.

Considering these species, the seeds are referred to as exalbuminous or nonendospermic. In these seeds, the stored reserves are said to be located in some parts of the embryo and this location is usually cotyledons.

Along with this, some of the species store reserves in nucellus. This is a maternal diploid tissue which is known as perisperm. There have been a number of books and many review papers that have been written on the endosperm. Some of them was written by Bhatnagar and Sawhney, 1981; Brink and Cooper, 1947; Vijayaraghavan and Prabhakar, 1984; Lopes and Larkins, 1993.

6.2. ENDOSPERM

Endosperm can be defined as a tissue that surrounds and nourishes the embryo in the seeds of angiosperms (that is the flowering plants). The endosperm is completely absorbed at maturity in some of the seed such as bean and pea. As it germinates, the fleshy food – storing cotyledons nourish the embryo.

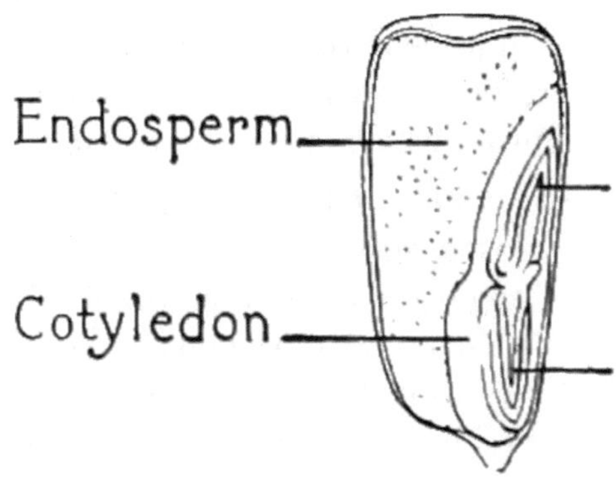

Figure 6.1: An image showing Endosperm and Cotyledon.

Source: Image by Flickr

There are some other seeds also, in which some of the endosperm is present until the process of germination and these include castor bean and wheat. Typically, the cotyledons are thin as well as membranous and moreover, serve to absorb the stored food from the endosperm while the process of germination takes place. Considering an example of coconut, the endosperm is liquidy. This liquid endosperm contains the substances which are important for growth. Endosperm is responsible for the economic importance of cereal grains as well as oilseeds.

The initiation of endosperm is considered to be a definitive characteristic of the double fertilization of angiosperms. The development of the same requires the fusion of at least one polar nucleus in the embryo sac along with one of the two sperm nuclei from the pollen grain. The nutritive material of the seed in gymnosperms is present prior to fertilization.

6.2.1. Origin and Function of the Endosperm Tissue

Intrinsically, the origin of endosperm is connected to the process of double fertilization. The double fertilization can be defined as a unique biological process in which the fusion of one sperm nucleus takes place with the egg in order to produce the embryo; whereas, the fusion of a second sperm nucleus takes place with two polar nuclei in order to form the triploid endosperm. These events tend to occur within the embryo sac. This is the sac which is embedded in maternal tissues (Dumas and Mogensen, 1993; Russell, 1993).

The term "*double fertilization*" refers to the fact that the primary endosperm nucleus is said to be derived from a fertilization event, despite the fact that the triple fusion nucleus does not go through a process which can lead to another organism, which is expected from true fertilization.

Sargant in the year 1900 firstly presented the discussion of the evolutionary origin of endosperm. Sargant was the person who hypothesized that in the ancestors of angiosperms, the double fertilization generated a second embryo. This is then evolved into a body along with storage function (the endosperm). This is the transformation which can have occurred due to the addition of a second female nucleus to the second fertilization event.

In the recent scenario, there have been a discovery that s second fertilization product produces embryo regularly in the non – flowering seed plant *Ephedra trifurca*. *Ephedra trifurca* is said to be one of the closest relatives to the angiosperm. This discovery provides its support for the first part of the hypothesis presented by Sargant (Friedman, 1992), despite the fact that the second sperm nucleus does not lead to a distinct nutritive tissue developmentally.

This has still been not known whether the endosperm tissue evolved before or after the addition of a third nucleus in order to form the second fertilization product. Originally, the endosperm can have been diploid, if a third nucleus is added to the second fertilization after the transformation of the supernumerary embryo into a nutritive tissue.

Nevertheless, it is important to be cautious while comparing this. The reason behind this is the fact that it is still not known whether double fertilization in flowering plants in Ephedra is evolutionarily homologous to double fertilization or not (Friedman, 1990b). The process of double fertilization can be important in order to preserve epigenetic differences to provide a direction to the development of the endosperm and embryo.

There was a hypothesis presented by Kermicle and Alleman (1990) that the gametic imprinting, which acts in conjunction with unequal gene contribution from the parents, could result in four effective gene dosage levels: *"a gene may be turned off when received from both parents, turned on following pollen transmission (one effective dose), turned on only following ovule transmission (two effective doses), or turned on when received from both parents (three doses)."*

Both were the authors who stressed that triploidy could be a mechanism of fine tuning effective gene dosage levels, under such condition. If the endosperm were diploid, then the imprinting would provide less advantage due to the fact that only three effective dosage levels would be possible.

Such a phenomenon or process can provide explanation relating it to a question that why the central cell does not distinguish and function as an endosperm, as it does the same work in gymnosperms and also, why the endosperm usually does not differentiate directly in sexually reproducing species from the sporophyte, as it does similar thing in autonomous apomitics (Koltunow, 1993).

As per the suggestions provided by Kermicle (in the year 1978), the generation of the epihybridity, or epistatic combinations of epigenetic variation can take place within the endosperm of self- pollinated and cross-pollinated species by differentiating the combination of depressed genes. If this is the case, then a novel form of hybrid vigor can be derived from epigenetic the differentiation between polar as well as sperm nuclei.

There can some other intriguing and unresolved aspects of endosperm origin and function. These aspects include its intermediary position between the old as well as the new sporophyte and the general need of a fertilization process for triggering its own development. Little or no reserve food is contained by the angiosperm ovule at the time of fertilization, in contrast to non - flowering plants. In non – flowering plants, a large female gametophyte tends to provide nutrients for the development of embryo.

The reduction of the female gametophyte in angiosperms has been provided by the evolution of an embryo – nourishing tissue from a supernumerary embryo (Friedman, 1992). According to Brink and Cooper (1947), the double fertilization may have provided the advantage of heterosis to the endosperm.

In addition to this, these were the authors who have suggested that its triploid nature can be considered as a way for acquiring the genetic advantage in an environment of diploid cells. Endoreduplication or the general polyploidization of the endosperm genome occurs during development and this can be viewed as a way for amplifying both advantages.

6.2.2. Various Types of Endosperm

Endosperm is said to be the name of food laden tissue and endosperm is made in order to nourish the embryo in seed plants.

Considering gymnosperms, it can be said that it shows the female gametophyte. In the case of angiosperms, the endosperm is considered to be a special tissue that is formed further resulting from vegetative fertilization, triple fusion or even fusion of a male gamete along with diploid secondary nucleus of the central cell. The fusion product is the primary endosperm cell which has a triploid endosperm nucleus.

Angiospermic endosperm can be divided into three types on the basis of the mode of its formation namely nuclear, cellular and helobial.

- Nuclear Endosperm:

Without wall formation, the primary endosperm nucleus tends to divide repeatedly in order to produce a large number of free nuclei. Along with this, the central vacuole tends to appear in the central cell and furthermore, pushes the cytoplasm which contains the nuclei to the periphery. The thickening of the cytoplasm takes place so as to decrease the size of the vacuole.

Eventually, it disappears along with the exception of a few cases. The multinucleate cytoplasm tends to undergo cleavage and also, it gives rise to a multi-cellular tissue for example, maize, rice, wheat, sunflower, *Capsella bursapastoris*. There are a number of cases, in which the cell wall formation remains incomplete.

Let's take an example of a coconut. The coconut has a multi-cellular endosperm referred to as coconut meal in the outer part. Also, there are free nuclear as well as vacuolated endosperms referred to as coconut milk in the center. Considering Pharsalus, it can be said that the wall formation occurs around the embryo only; whereas, in Crotalaria, it tends to be restricted to the upper half.

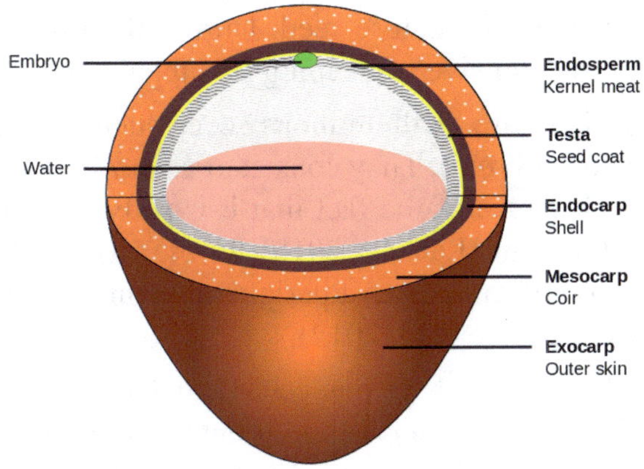

Figure 6.2: Coconut layers.

Source: Image by Wikimedia Commons

Nuclear endosperm is said to be the most common type of endosperm. The name is given because of the fact that it contains free nuclei in the beginning.

- Cellular Endosperm:

In the primary endosperm nucleus, every division is followed by cytokinesis. Thus, the endosperm becomes cellular from the starting for example, Datura, Balsam, Petunia.

- Helobial Endosperm:

Helobial Endosperm tends to occur in the order Helobiales of monocots. The endosperm is said to be of an intermediate type in between the cellular type and the nuclear type. In the primary endosperm nucleus, the first division is followed by transverse cytokinesis in order to form two unequal cells namely, larger micropylar and smaller chalazal. The micropylar cell tend grow faster in comparison to the chalazal one.

In both the cells, further development occurs similar to that of nuclear endosperm, which is the multinucleate stage followed by wall formation. An example can be taken of Asphodelus. Nevertheless, most of the time the chalazal chamber remains smaller and can even degenerate.

6.2.3. Structural diversity of the endosperm in mature seeds

In mature seeds, the variation in the structure of endosperm can take place considerably between different species. Let's take an example of tomato: named Solanum *lycopersicum*. The tomato consists of a hard and thick endosperm cell layer in the mature seed. During the time of germination, this tends to undergo extensive weakening (Nonogaki et al. 2000).

Contrastingly, there are exalbuminous seeds such as soybean (*Glycine max*) and pea (Pisum sativum). They consist of a very little or no endosperm in the mature seed because of the fact that it is fully consumed during the time of seed development (Lackey. 2010). In the case of Arabidopsis, the endosperm is said to be confined to a peripheral aleurone- like -cell layer in the mature seed (K.J.D. Lee et al. 2012).

Moreover, it tends to act as a mechanical barrier for inhibiting the embryonic growth, and also, it acts as a nutrient reserve for seed germination as well as early seedling establishment. The seed coat tends to undergo PCD that is programmed cell death during the time of seed development. As a result of this, it is no longer alive in the mature seed.

In opposite to this, the cereal seeds have much more complex structures along with a large starchy endosperm as well as living aleurone layers. Considering the most cereal grains, the aleurone is said to be a single cell layer. But there is an exceptional case also and that is of the barley (*Hordeum vulgare*) grain. Barley consists of three cell layers.

6.3. DEVELOPMENT OF THE ENDOSPERM

6.3.1. Initiation of endosperm development

The characterization of the plant reproduction is done by a double fertilization which further results in the production of the embryo and the endosperm. Considering plants, the male and female gametes are produced after the process of meiosis and follow a series of divisions of a haploid male or female spore. Male gametogenesis in angiosperms takes place in stamens and furthermore, results in pollen which contains two sperm cells.

The female gametogenesis tends to take place within the diploid tissues of the ovule. The production of the haploid megaspore is done by meiosis and this undergoes a series of three syncytial divisions. Furthermore, it is followed by cellularization and so, produces the embryo sac. The embryo sac consists of the haploid female gamete or egg cell and the central cell.

The pollen tube tends to deliver two sperm cells to the female gametophyte. One of the female gametophyte tends to fuse with the egg cell and one that fuses with the central cell. The fertilization of the egg cells that result in embryogenesis.

The production of the endosperm takes place by the fusion of the central cell along with the other sperm cell. There are a number of plant species, in which the central cell inherits two haploid nuclei from the syncytial female gametophyte.

Therefore, the endosperm genome consists of two doses of the maternal genome and one dose of the paternal genome. There can be variations in the ploidy of the endosperm amongst the angiosperms. This can be a result of a varied dosage of the maternal genome.

Therefore, it is important to focus solely on endosperm development in *A. thaliana* and cereals. There are some extensive reviews which provide details of other modes of development and these reviews include Floyd & Friedman, 2000; Maheshwari, 1950; Olsen, 2001.

6.3.2. Modes of endosperm development

The endosperm development tends to follow three major phases. Once the fertilization is done, the central cell immediately undergoes the mitoses which are further not followed by the cell division.

This tends to result in the syncytial endosperm developmental phase. This is the phase that ends with the partition of the multi-nucleated syncytial endosperm into the individual cells.

This is followed by the cellular phase. During this cellular phase, the differentiation of the cells divide and the various cell types takes place. The ultimate maturation phase of endosperm development is marked by an arrest of cell division along with a change of metabolism.

This results in reserving the accumulation in the seed. There can be variation seen amongst the species in the duration and importance of these three phases.

- Syncytial phase

This mode named syncytial phase is the mode of syncytial division. This mode is typical of large cells which are rapidly growing. This is observed in some other tissues which is related to the nutrient transport function such as feeding cells induced by parasitic nematodes or the vascular system of plants (Favery et al., 2004) and there can be some cells in the mammalian placenta (Cross, 2005).

Considering A. *thaliana*, there are a series of eight syncytial divisions which result in 200 nuclei. This is mostly located at the periphery of the endosperm (Mansfield & Briarty, 1990; Brown et al., 1999; Mansfield & Briarty, 1990; Boisnard-Lorig et al., 2001). The semi - synchronicity of the syncytial divisions allows the classification of the syncytial phase into successive stages I to VIII (Ingouff et al., 2005).

Figure 6.3: An image of *Arabidopsis thaliana* inflorescencias.

Source: Image by Wikimedia Commons

A single large vacuole occupies the center of the syncytium. Each and every nucleus is further surrounded by a mass of cytoplasm. This is a boundary which is delimited by a dense cortical array of cytoskeleton. Such kind of a unit has been defined as NCD that is nucleo-cytoplasmic domain (Brown et al., 1999). For the first three syncytial divisions, the pace of cell division is c. 2–3 h per cell cycle and during subsequent cycles, it tends to gradually increase to 12 h.

The first division results in eight endosperm nuclei and they are evenly spaced along a curved tube bounded by the inner seed integument (Boisnard-Lorig et al., 2001; Ingouff et al., 2005). During the following cycle of syncytial division, there are nuclei at the chalazal pole opposite to the site of sperm entry, which do not tend to enter a further round of mitosis. They presumably undergo endoreduplication as their enlargement take place.

The larger nuclei are the first markers of the posterior endosperm pole. This is also termed the CZE that is chalazal endosperm. In the syncytial division, the next round is marked by the definition of a second mitotic domain at the anterior pole that consists of five to eight nuclei surrounding the embryo.

The direct dynamic observation of mitosis and the regional expression of the mitotic cyclins has been done. This observation shows that the division of these nuclei is done earlier to the nuclei in the peripheral mitotic domain. After that, an independent pace is maintained by the cell cycle in each of the three domains.

Furthermore, the divisions in the peripheral domain result in stages VII (c. 50 nuclei), VIII (100 nuclei) and IX (200 nuclei). In cereal species, the mode of syncytial endosperm development seems to be much similar to that is described above and also, the details of which has been provided by Brown & Lemmon (2007).

- Cellularization

In the nuclear division, the eighth cycle marks the ends of the syncytial phase due to the fact that it if followed by the cellularization in the anterior as well as peripheral domains. This results in the isolation of most nuclei in individual cells. Considering A. *thaliana* time-lapse recordings, it has been shown that the endosperm cellularization tends to occur immediately after the eighth mitotic division (Sorensen et al., 2002).

In the mutants that are affected in the basic mechanisms of cytokinesis (which are also active in the embryo), cellularization is not present (Sorensen et al., 2002). In mutants (those who have reduced the endosperm growth), the cellularization occurs early. Nevertheless, the pace of nuclear division tends to remain unaffected and this in turn results in a decreased number of cells in the mutant cellular endosperm (Garcia et al., 2003).

This provides suggestions that a critical nucleo-cytoplasmic can show a signal, which tends to trigger cellularization. An important role is played by the type I MADS box transcription regulator AGAMOUSLIKE 62 (AGL62) in the control of cellularization (Kang et al., 2008). The expression of AGL62 is confined to the endosperm after the occurrence of fertilization and tends to drop abruptly before the occurrence of cellularization.

There can be mutants deprived of AGL62 and they can show very precocious cellularization, and further suggests that AGL62 provides prevention to the onset of cellularization during the process of syncytial endosperm development. The process of cellularization itself is said to be a variation of conventional cytokinesis along with notable differences in cytoskeletal arrangements (Otegui & Staehelin, 2000a; Olsen, 2001; Otegui et al., 2001).

Most of the time, the endosperm cellularization relies on the functions important for cytokinesis in vegetative cells (such as actin and associated proteins, microtubules, and vesicle trafficking) (Sorensen et al., 2002). The formins act as actin polymerization-regulating factors (Blanchoin & Staiger, 2010; Bugyi & Carlier, 2010; Campellone & Welch, 2010) and therefore, play an important role in endosperm cellularization (Ingouff et al., 2005a; Fitz Gerald et al., 2009).

Moreover, this suggests that a role is played by actin cytoskeleton also cellularization, similar to the role which it plays in the plant cytokinesis. Therefore, it is much likely that endosperm cellularization majorly relies on the same molecular components as in the plant cytokinesis. Nevertheless, the formation of cell walls around syncytial NCDs probably depends on distinct architectural organization of such conserved components.

Each NCD is delimited which is done by a dense array of microtubules further defining the future division sites. No preprophase band tends to anticipate the spatial position of the new cell wall during the process cellularization and this is not similar to that seen in plant cytokinesis (Otegui & Staehelin, 2000b).

After the process of the critical last syncytial mitosis is done, the formation of a series of cell plates takes place between the sister nuclei that can said to be as a result of a rather complex mechanism based on vesicles. These are the vesicles derived from the endoplasmic reticulum (Otegui et al., 2001). The cell plates get assembled in between non-sister nuclei while defining boundaries can be either hexagonal or pentagonal shaped around each NCD.

At the same time, the endosperm cell plates arise around all NCDs (Sorensen et al., 2002). This further implies that some coordination mechanisms, that is not known, make sure that each and every cell plate makes connection only once to its multiple neighbors. This connection forms three - way junctions, and that is not so common in conventional cytokinesis.

There is one more idiosyncratic feature of cellularization and this feature lies in the centripetal growth of cell plates (Olsen, 2001). After the occurrence of the cellularization, there are a number of rounds of cell division which take place. The formation of cell plates takes place from the center to the periphery in all the directions, similar to that during conventional cytokinesis (Olsen, 2001, 2004; Nguyen et al., 2002).

6.3.3. Cellular phase and maturation

After the occurrence of cellularization, the cellular developmental phase involves a small series of synchronous cell divisions that results in the formation of four to six layers of endosperm cells. This further tends to depend on the position along the MC that is micropylar–chalazal axis.

The division takes place in the central part of the peripheral endosperm and moreover, it progresses towards the chalazal pole; whereas the central vacuole recesses in a progressive manner until the whole peripheral endosperm becomes cellular.

Considering A. *thaliana* and a number of dicotyledonous species, it can be said that the endosperm does not persist since they produce seeds with morphology in comparison to that of legumes. There exists an exception of the cell layer immediately when it comes in contact with the seed coat.

There are functions of this layer connected to the control of seed dormancy. Its function has been reviewed thoroughly and the examples include Koornneef et al., 2002; Finkelstein et al., 2008; Finch-Savage & Leubner-Metzger, 2006; Angelovici et al., 2010; Raghavendra et al., 2010. In the embryo, the reserves are stored and this proliferates rapidly at the expense of the cellular endosperm and this happens after the heart stage. The synthesis of starch takes place in the cellular endosperm and amyloplasts distinguish.

There has been limited attention received by this phase in *A. thaliana* (Mansfield & Briarty, 1990). There have been recent investigations in some other species and these investigations have provided important physiological insights. In the case of cereals, the maturation of the endosperm is marked through endoreguplication as well as programmed cell death.

The endoreduplication tends to coincide with the collection of reserves in the starchy endosperm, despite the fact that in some of the species, the aleurone cells can also endoreduplicate. The endoreduplication is gradual and it also tends to be involved probably along with the RBR3 that is RETINOBLASTOMA RELATED 3. This controls the G1 / S transition (Sabelli et al., 2005; Nguyen et al., 2007).

Similar to the other seed components, the development of the endosperm ends with dessication. The starchy endosperms get affected by the programmed cell death in cereals. It goes in sync with the peaks of ethylene and the elevated concentrations of abscisic acid. There can be a possibility that most of the endosperm cells which are loaded with reserves can die, along with the exception of the outer cell layer, which is referred to as the aleurone layer in cereals (Sabelli & Larkins, 2009).

6.3.4. Varied modes of endosperm development

Despite the fact, as per the descriptions provided from *A. thaliana* and cereals, the sequence of the endosperm development have been observed in a number of other plant genera. There can be a variation in the duration and importance of the syncytial as well as cellular phases of development. These variations can vary from one species to another, which in turn results in extreme cases of pure syncytial endosperm development or pure cellular development.

There is the initial division of the central cell into two compartments that is then followed by syncytial division in each or only one of the compartments. These have also been described as 'helobial' endosperm development (Maheshwari, 1950; Vijayaraghavan & Prabhakar, 1984). There have also most of the basal angiosperms which show the prevalence of cellular endosperm development.

The initial division of the central cell into two compartments followed by syncytial division in each or only one of the compartments has also been described as 'helobial' endosperm development. (Maheshwari, 1950; Vijayaraghavan & Prabhakar, 1984). Most basal angiosperms show the prevalence of cellular endosperm development (Floyd & Friedman, 2000, 2001).

Therefore, it is considered that the syncytial endosperm development as well as storage of reserves in the embryo show more steps of endosperm evolution keeping in mind an ancestral helobial or cellular mode (Floyd & Friedman, 2000, 2001).

6.3.5. Endosperm development and its role in seed development

The cells of the endosperm are triploid (3n) and these cells tend to arise from the fusion of polar nuclei (2n) along with the sperm nucleus (n) by a double fertilization event. Among the plant species, the development of the endosperm is divergent. However, some similar features have been represented while comparing the endosperm development between cereals and Arabidopsis (Arabidopsis thaliana) (Olsen, 2004).

In such kind of plants, the development of the endosperm can be further divided into different phases. These includes the formation of the nuclear endosperm (or coenocyte-type endosperm), cellularization, differentiation, maturation and cell death. While, developing the nuclear endosperm it includes repetitive nuclear division without separating the nuclei by the cell wall.

This process is then followed by the process of cellularization. This separates sister nuclei by forming a periclinal cell wall. In the cereal endosperm, there are majorly four cell types namely the starchy endosperm, transfer cells, the aleurone layer, and the region which surrounds the embryo.

The starchy endosperm refers to the major tissue for the purpose of seed reserve accumulation in the cereal grains. This in turn serves as a nutrient source in the process of seed germination and seedling establishment.

Contrasting to cereals, in dicot seeds, the cellularization occurs in the endosperm and then the endosperm is consumed to be used as a source of energy for the growth of embryo during the seed development (Lopes and Larkins. 1993).

Small mature dry seeds are produced by the Arabidopsis mutants defective in endosperm cellularization and this is done due to the lack of nutrients that is required for the growth of embryo (Sørensen et al. 2002).

The radicle refers to the part which protrudes during the process of seed germination. There is an endosperm that is adjacent to the radical and this is referred to as ME or micropylar endosperm. The ME (micropylar endosperm) tends to express a set of endosperm - specific genes. These genes include those which are related to cell wall loosening (Dekkers et al. 2013).

The visualization of the ME - specific gene expression has been done by a number of techniques. These techniques include tissue print (Nonogaki et al. 2000), enhancer trap in Arabidopsis (Liu et al. 2005) and *in situ* hybridization (Iglesias-Fernandez et al. 2013).

6.4. RELATIONSHIPS BETWEEN ENDOSPERM, EMBRYO, AND MATERNAL TISSUES

There is a genetic balance in between the endosperm, embryo and the maternal tissue which has been long known to be a key requirement for normal seed development. Considering the maize plants that carry the indeterminate gametophyte (ig) mutation, it can be said that the accessory polar nuclei are frequently present.

Once the pollination done, the generation of endosperms with ploidy levels takes place, which ranges from diploid to octaploid along with only one chromosome set of paternal origin. However, full development of only the triploid endosperm with the standard ratio of one paternal to two maternal genomes takes place (Lin, 1984).

Along with this, there were genetic imbalances which were generated by enforced self – fertilization, crossing species which tend to vary in ploidy level and aneuploidy. These genetic imbalances can result in severe developmental defects in the endosperm and also, lead to seed abortion in a number of species (Brink and Cooper, 1947). This kind of interactions between endosperm, embryo and maternal tissues became one of the most complex as well as unresolved aspects of seed development.

There have been differences in the activity of maternally as well as paternally derived genomes in maize endosperm (Kermicle and Alleman, 1990). These differences shows that there can be possibility that a number of aspects of endosperm development and the relation possessed by the endosperm and the embryo and maternal tissues are defined by genes which are differentially imprinted.

Nevertheless, it is important to have a better understanding of the mechanism of gametic imprinting along with its precise role in seed development. The present knowledge that humans have are related to the complex developmental interactions which are taking place in a number of living organisms. Thus, the repertoire of morphogenetic events results in the progressive development of endosperm and embryo within maternal tissues which will eventually challenge the developmental biologists for many upcoming years.

The isolation and characterization of mutations which disrupt the normal development have proved to be useful while addressing a number of different aspects of endosperm development. This includes the complex interactions which take place among endosperm, embryo and maternal tissues. As per the analysis done by Neuffer and Sheridan (in the year 1980), it can be seen that 855 maize ethylmethane sulfonate-induced defective kernel (dek) mutants represent approximately 285 different loci and are scattered throughout the genome.

Among these, there were many loci which have showed effects on both embryo as well as endosperm and there were a very few which could be assigned potentially to endosperm – specific genes. There were similar results which were reported for defective kernel (sex) mutants while considering the case of barley (Bosnes et al. 1987).

Both were the studies that suggested that a large set of seed – specific genes is expressed in both endosperm as well as embryo, despite the fact that these tissues are remarkably different while comparing it at morphological as well as physiological levels.

The conclusion of such reports has been a support for the hypothesis done by Friedman (1990, 1992). The hypothesis was related to the endosperm evolution from a supernumerary embryo (Olsen et al. 1992).

The genetic analysis provides suggestions that the maternal and endosperm tissues can regulate the development of each other. Considering the case of cereals, the base of the endosperm distinguishes into a number of different layers of transfer cells along with extensive wall ingrowths.

This is the specialized region which functions in metabolite transport and furthermore, it provides an indirect connection with the maternal tissues. The process of protein synthesis and the process of starch biosynthesis are majorly dependent on the metabolic events which take place in this region. However, there has been a little research which has been directed on its biochemical as well as molecular characterization.

The maize mutant miniature - 7 (mnl) was first described by Lowe and Nelson in the year 1946. This in turn provides illustration about the important role of the basal endosperm cells and their interaction with the maternal tissue that are present nearby. As per Miller and Chourey (1992), the developing seeds of this mutant have decreased drastically invertase activity.

Among the sucrose which tends to arrive at the base of the endosperm from maternal cells, most of that is cleaved to glucose as well as fructose by invertases. Then, these two hexodes move to the other parts of the endosperm.

Here, the sucrose is partly synthesized again and moreover, is used (Shannon et al., 1986). In mn7 mutants, at the base of the endosperm, the loss of invertase activity results in the destruction of maternal cells which transport the sucrose.

Therefore, the endosperm tends to fail while accumulating the starch and protein (Miller and Chourey, 1992). On the flipped side, when the analysis of the maternal effects in shrunken endosperm (seg) barley mutants is done then it suggests that the maternal transcripts could mediate endosperm development (Felker et al., 1985).

This is considered to be much interesting and the reason is that the gene products which are maternally coded have been shown to play important roles in the early morphogenetic events. This in turn results in specification of the body axes in *Drosophila* larvae (Ingham, 1988).

It is important to understand the biological cross-talk because it allows the endosperm genome to control the phenotype of a maternal tissue and vice versa. This tends to put light on some of the complex interactions that takes place in between these tissues.

In ploidy level and genetic constitution, the differences showed in the relationship between the triad endosperm – embryo - mother plant, results in a number of questions which are related to the integrated development of these tissues, especially considering the relation to nutrient flow.

The flow of nutrients takes place from 2n maternal tissues to a 3n endosperm and a 2n embryo. In these three tissues, the ratio of maternal to paternal constitution is 2:0, 2:1, and 1:1 respectively. It is still not known that whether the regulation of the nutrient flow is done by coordinated genome action or whether each and every phase is under the independent control of a respective genome (Evenari, 1984).

The process of vascularization tends to stop prior to the point at which the connection between the maternal tissues and the developing endosperm occurs. There can be some questions which can arise like 'does this indicate a boundary for controlling the nutrient flow?

Considering the case of cereals, it can be said that there is an apparent lack of symplastic connections in between vegetative plant parts and the endosperm (Felker and Shannon, 1980; Thorne 1980, 1985) and this makes it difficult for finding the obvious physiological mechanisms which are involved in the flow of nutrients from maternal tissues to the growing seed.

Despite the fact that there is evidence that exist for transporting the apoplastic nutrient from maternal tissues to the cells of cereal endosperm (Jenner, 1974; Shannon et al. 1986), there can be absence of similar evidence for most of the species and along with this there can be poor understanding of gene activity in the regions which are involved in nutrient transport which can create difficulty while identifying the key steps while regulating this complex process.

Along with its better known function is to act as a source of nitrogen, sulfur, minerals and energy for the embryo during the germination process. Also, the endosperm is generally considered to be a medium which provide support to the embryogenesis. There is not much which is known regarding the time at which the embryo starts in order to utilize the endosperm for nutrition.

It is still not clear whether the embryo receives nutrients passively which are mobilized by the endosperm or whether coordination is there in such mobilization by further developing the embryo itself or by a combined action of both the tissues.

There can be structural and physiological adaptations that can control such kind of interactions and furthermore, balance the growth of embryo and endosperm. It is important to have such a balance but again there is no clue about how establishment and regulation of these interactions take place.

In a number of species, the mitotic activity starts earlier in the primary endosperm nucleus while comparing it with the zygote. This fact could indicate an early dependency of the embryo on endosperm development (Marinos, 1970; Raghavan, 1986). Nevertheless, it is still not possible to judge the way in which this interaction is established as well as regulated.

In a number of angiosperms, the ephemeral nature of the endosperm tissue is an intriguing aspect but it is largely ignored when it comes to seed development and evolution.

Considering many species, there can be absence of endosperm in mature seeds and this is known to produce it at early stages of ovule development. This seems to provide indications that the consumption of the endosperms is done by the growing embryo (Raghavan, 1986).

This process is considered to be extreme in a number of dicotyledonous species and in turn assimilate the endosperm during the process of embryogenesis. Along with this, they tend to produce embryonic leaves (that is cotyledons) which act as a means of storing nutrient.

There are one of the few examples of cytological relationships between the developing embryo and endosperm and the mechanisms which are involved in the embryo nutrition which were provided by Marinos (in the year 1970).

Figure 6.4: Germination Dicotyledon Sprout.

Source: Image by Pixabay

Marinos studied the developing pea seeds which in turn showed that the absorption of endosperm is done in the free – nuclear state prior to the beginning of cell walls formation. Along with this, the acellulosic wall structure tends to appear in the endospermic cytoplasm. Furthermore, Marinos suggested that this can get involved in the precise positioning of the embryo in the embryo sac and controlling the transfer of nutrient to the embryo.

In the similar manner, the case of sunflower (Newcomb and Steeves, 1971), cotton (Schulz and Jensen, 1977), and alfalfa (Sangduen et al., 1983) is considered. In them, the inner wall of the embryo sac tends to produce wall invaginations furthermore appearing to be the specialized transfer structure into the endosperm. There is no information regarding the physiological mechanisms which are involved in the process of endosperm degradation during the process of embryogenesis.

There were questions asked by Evenari (1984) in relation with this that remained majorly unanswered. The questions were:

- If the endosperm is digested by the embryo, what kind of signal triggers this process?

- Does the embryo digest the endosperm? Or
- Does it just absorb its contents after they are made available by some process of endosperm degeneration?

Considering the evolutionary side, it is not obvious that any kind of advantage would be provided by such remobilization. Dure (1975) pointed out that, "The detailed biochemistry of this nutrient flow, with its rise and fall of tissues, its integration and regulation, presents a unique and potentially rewarding system for developmental biologists."

There are a number of variations which can take place in embryo and endosperm genetic constitution and these variations are generated by apomixis. Apomixis can be defined as a developmentally instructive deviation in sexual reproduction (Koltunow, 1993). Asexual seed formation is contained in the apomixes which in turn results in offspring with a maternal genotype. Furthermore, it allows for the maintenance of alteration of generations without even affecting the genotype.

The autonomous apomicts are common in the families, Asteraceae and Compositae. In this, the development of embryo and endosperm takes place independently of pollination. Considering the pseudogamous apomicts, it can be said that the fertilization of the polar nuclei by a sperm nucleus tends to appear to be important in order to initiate the endosperm development, despite the fact that the formation of the embryo is done from an unfertilized egg cell.

In a number of species, the existence of adventitious embryogenesis is there, with no formation of embryo sac and related endosperm (Ramachandran and Raghavan, 1992; Koltunow, 1993). This further indicates that under appropriate conditions, the somatic cells have the capacity to express a developmental program which results in egg cell and embryo formation (Goldberg et al., 1989). Also, it indicates that there is no absolute importance for the endosperm to act as an accessory tissue for normal embryogenesis.

6.5. FUTURE PROSPECTS IN ENDOSPERM RESEARCH

There is not much known of the biochemical composition of the endosperm tissue of a number of species. There are components including storage and non – storage proteins, oils, phytin, carotenoids, free amino acids, polysaccharides and phenolics that are known to exist in various proportions in the endosperm.

When there is a better understanding about their functional roles, cellular locations and effects on nutritional quality, germination and seed storage, then this will provide vast knowledge for improving the seed quality and their use. In cereals, the improvement can be done in the nutritional quality. This type of improvement has been hampered by a lack of knowledge of endosperm proteins that consists of nutritionally desirable amino acids.

The cereal cultivars having high nutritional quality have been developed traditionally on the basis of the use of mutations that can even have undesirable pleiotropic effects. When the development of biochemical assays is done, it allows the efficient characterization of endosperm components among a large number of progeny to help breeders in exploring the existing genetic variation to develop the improved cultivars.

Various genes particularly expressed in endosperm can allow the development of efficient reporter gene systems. The transformation techniques have been said to be useful in order to probe gene regulation and cell function.

For a number of years, the research has been done in this area which will likely to be important. The reason is that the identification of endosperm – specific promoters is required in order to develop the engineered genes that can result in the accumulation of novel or quality – enhanced products in the endosperm. The molecular biology is already providing alternative and complementary approaches to the conventional breeding techniques so as to improve the plants.

The encoding of the expression of high levels of nutritionally desirable proteins is done by introduced genes and these will ultimately result in the development of seed crops having improved nutritional quality. The increase in yield mainly relies on the possibility of increasing the efficiency of deposition of min endosperm reserves like starch.

Let's take an example. The transgenic potatoes contain a mutant *Escherichia coli* gene encoding ADP-glucose pyrophosphorylase. This tends to enhance a rate – limiting step in starch biosynthesis. These transgenic potatoes have a 30 percent increase in starch content in the tuber (Stark et al., 1992). When this is achieved, it will provide hope that same developments having engineered cereal endosperm will lead to cultivars having higher yield potential.

There are a number of aspects of endosperm development, cell biology, physiology and biochemistry considering that it can be said that the endosperm is unique. It has an apparent simplicity which is reflected by the presence of a few cell types which tend to differ from each other. It has permitted detailed studies on differentiation and the biosynthesis of some specific stored metabolites.

Nevertheless, there has been a little attention given to understanding the genetic as well as developmental basis of endosperm differentiation and the function. Considering the time in the past, the apparent morphological simplicity of endosperm tissue resulted in many so as to view it as a mass of storage cells or a simple appendage to the embryo.

There are some people who believe that the flow of nutrients from maternal tissues can be considered as the only factor which guides differentiation in this tissue. However, in cereal endosperms, the characterization of developmental mutants is starting to show that the complex genetic controls are responsible for the remarkable constancy which can be observed while developing the endosperm tissue.

For research purposes, the study of interactions of endosperm, embryo, and maternal tissues is an exciting area. This area has the ability to provide answers to a number of questions which are related to the complex temporal as well as spatial interactions which affect the development of plant tissues.

Still, it is surprising that this area gets so little attention to the study of the developmental differences between species in which the absorption of the endosperm takes place during the process of seed development versus species in which absorption of this tissue takes place during the process of germination. In addition to this, there are still many questions which are left unanswered regarding the regulation of endosperm digestion during the process of germination of endospermic species, which lack an aleurone.

The double fertilization holds the functional importance in angiosperms and this important is considered as one of the questions which are most puzzling in plant biology.

There has been evidence that in maize, at least, the products of double fertilization are differentially imprinted points probably that the epigenetic differences in progenitor nuclei can be considered responsible for the steps that tend to result in endosperm and embryo differentiation (Kermicle and Alleman, 1990).

6.6. CONCLUSION

In the case of angiosperms, there is a process named double fertilization event that tends to initiate the development of two distinct structures that includes the embryo and the endosperm. An important role is being played by the endosperm while providing support to the embryonic growth. This is done by supplying nutrients, protecting the embryo and controlling embryo growth and here, it tends to act as a mechanical barrier during the process of seed development and germination.

The chapter includes the structure and function in the mature dry seed that is said to be divergent and specialized among a number of various plant species. A subset of endospermic tissues contains living cells even after the process of seed maturation is done, and thus, play an active role while regulating the process of seed germination.

By reading this chapter, readers will get to know in detail that the embryo secretes signals to the endosperm in order to induce the degradation of the seed reserve and for promoting the endosperm weakening during the process of germination.

The conclusion that can be drawn from the chapter is that there have recent advancements in seed biology that have shown that the endosperm is capable of sensing environmental signals, and moreover, it can produce and secrete signals in order to regulate the growth of the embryo. Therefore, the process germination is said to be a systemic response that involves bidirectional interactions between the embryo and endosperm.

REFERENCES

1. Demason, D., n.d. *Endosperm Structure And Development*. [ebook] p.1. Available at: <https://link.springer.com/chapter/10.1007%2F978-94-015-8909-3_3> [Accessed 15 May 2020].
2. Encyclopedia Britannica. 1998. *Endosperm | Definition, Description, & Importance*. [online] Available at: <https://www.britannica.com/science/endosperm> [Accessed 15 May 2020].
3. Li, J. and Berger, F., 2012. Endosperm: food for humankind and fodder for scientific discoveries. *New Phytologist*, [online] 195(2), pp.290-305. Available at: <https://nph.onlinelibrary.wiley.com/doi/pdf/10.1111/j.1469-8137.2012.04182.x> [Accessed 15 May 2020].
4. Lopes, M. and Larkins, B., 1993. Endosperm origin, development, and function. *The Plant Cell*, [online] 5(10), pp.1383-1399. Available at: <https://www.researchgate.net/publication/14918334_Endosperm_Origin_Development_and_Function> [Accessed 15 May 2020].
5. Lopes, M. and Larkins, B., 1993. *Endosperm Origin, Development, And Function*. [ebook] p.2. Available at: <http://www.plantcell.org/content/plantcell/5/10/1383.full.pdf> [Accessed 15 May 2020].
6. Mondal, P., n.d. *Various Types Of Endosperm Of Flowering Plants*. [online] Your Article Library. Available at: <http://www.yourarticlelibrary.com/biology/various-types-of-endosperm-of-flowering-plants/11785> [Accessed 15 May 2020].
7. Yan, D., Duermeyer, L., Leoveanu, C. and Nambara, E., 2014. The Functions of the Endosperm During Seed Germination. *Plant and Cell Physiology*, [online] 55(9), pp.1521-1533. Available at: <https://academic.oup.com/pcp/article/55/9/1521/2756388> [Accessed 15 May 2020].
8. Yan, D., Duermeyer, L., Leoveanu, C. and Nambara, E., 2014. The Functions of the Endosperm During Seed Germination. *Plant and Cell Physiology*, [online] 55(9), pp.1521-1533. Available at: <https://www.ncbi.nlm.nih.gov/pubmed/24964910> [Accessed 15 May 2020].
9. Amara et al. 2012
10. Angelovici et al., 2010
11. Bai et al. 2012
12. Blanchoin & Staiger, 2010
13. Boisnard-Lorig et al., 2001

14. Bosnes et al., 1987
15. Brink and Cooper, 1947
16. Brown & Lemmon, 2007
17. Brown et al., 1999
18. Bugyi & Carlier, 2010
19. Campellone & Welch, 2010
20. Cross, 2005
21. Dekkers et al. 2013
22. Dumas and Mogensen, 1993
23. Esau, 1977
24. Evenari, 1984
25. Favery et al., 2004
26. Felker and Shannon, 1980
27. Felker et al., 1985
28. Fernández-Arbaizar et al. 2012
29. Finch-Savage & Leubner-Metzger, 2006
30. Finkelstein et al., 2008
31. Fitz Gerald et al., 2009
32. Floyd & Friedman, 2000
33. Friedman, 1992
34. Garcia et al., 2003
35. Goldberg et al., 1989
36. Iglesias-Fernandez et al. 2013
37. Ingham, 1988
38. Ingouff et al., 2005
39. Jenner, 1974
40. K.J.D. Lee et al. 2012
41. Kang et al., 2008
42. Kanno et al. 2010
43. Karssen et al. 1983
44. Kermicle and Alleman, 1990
45. Kermicle, 1978
46. Kim et al. 2013

47. Koltunow, 1993
48. Koornneef et al., 2002
49. Lackey 2010
50. Lee et al. 2010
51. Leymarie et al. 2012
52. Li et al. 2013
53. Lin, 1984
54. Liu et al. 2005
55. Maheshwari, 1950
56. Mansfield & Briarty, 1990
57. Marinos, 1970
58. Martinez-Andujar et al. 2012
59. Matakiadis et al. 2009
60. Miller and Chourey, 1992
61. Muller et al. 2006
62. Nambara and Nonogaki 2012
63. Nambara et al. 2010
64. Newcomb and Steeves, 1971
65. Nguyen et al., 2007
66. Nonogaki, 2010
67. Oh et al. 2007
68. Olsen, 2001
69. Otegui & Staehelin, 2000a
70. Otegui et al., 2001
71. Paine et al. 2005
72. Penfield et al. 2006
73. Preston et al. 2009
74. Raghavan, 1986
75. Raghavendra et al., 2010
76. Ramachandran and Raghavan, 1992
77. Russell, 1993
78. Sabelli & Larkins, 2009
79. Sabelli et al., 2005

80. Sangduen et al., 1983
81. Sargant, 1900
82. Schulz and Jensen, 1977
83. Seo et al. 2009
84. Shannon et al. 1986
85. Sorensen et al., 2002
86. Stark et al., 1992
87. Thorne 1980, 1985
88. Toh et al. 2012
89. Tuncel et al. 2014
90. Vijayaraghavan & Prabhakar, 1984
91. Wang et al. 2013
92. Yamauchi et al. 2004

Chapter 7

Seed Dormancy and Germination

CONTENTS

Seed dormancy permits seeds to overcome unfavorable periods for the formation of seedlings and therefore, would be considered vital for plant ecology and agriculture. The induction of dormancy and the shifting of the dormant to germinating state, numerous processes are supposed to be involved.

The seed dormancy would play an important part in adaptation and evolution of seed plants. Though it would have a clear biological significance, the molecular mechanisms encompassing induction, maintenance and release of seed dormancy would still be considered ambiguous. For investigating gibberellin and abscisic acid metabolism in seeds, serious efforts are considered to have been implemented, hugely contributing to the prevailing understanding of the mechanisms of seed dormancy.

The chapter describes the role of plant hormones, the various tissues as well as various tissues and genes engaged which also comprises of the genes that have been newly identified, in regards to dormancy and germination. Additionally, the chapter also highlights the various analyses like transcriptome, proteome, etc. for getting better knowledge of such mechanically complex processes.

7.1. INTRODUCTION

Reproduction is considered to be an important stage in the life cycle of a plant. Hence, the genes that affect the seed dormancy and germination in natural plant population, would be the ones that fall in the stronger selection category. Seed dispersal would be terminated by germination to impact the place and time of plant growth. Moreover, germination could be interrupted through seed dormancy.

In reality, seeds would generally be considered as dormant or non-dormant. However, the ones with highly stimulating dormancy-inducing pathways, would be germinating in a consistently limited environmental ambit.

Thus, the dormancy measurements and environmental context analysis are supposed to be constantly coexisting, with a description of phenotypes or behavior being made. In simpler terms, by forming a basic physical wall around the seed so as to obstruct any king of exchange of gases or passage of water, dormancy could be established.

The so-called ‘physical dormancy’ in seeds would often need either an opening on the physical wall of the seed or movement through an animal gut (which would be full of digestive enzymes). It would result in the disintegration of the wall and allow germination. In 'morphological dormancy', the embryo inside the seed would remain nascent at maturity along with a prevailing dormant stage by means of which growth of the embryo would progress after shedding the physical wall. Finally, it would result in destruction of the tissues surrounding it.

Figure 7.1: Seed Germination.

Source: Image by Wikimedia Commons

Significantly, most of the seeds would show 'physiological dormancy'. It is an inactivity program started by the embryo or the encompassing endosperm tissues. In physiological dormancy, to restrict germination in a situation where there would be no particular environmental stimulates for its development, germination-inhibiting hormones would be used.

During and after germination, by the absorption of stored reserves of protein, oil, or starch collected at the time of seed maturation, early seedling growth is supported. Moreover, till the time photo auxotrophic activity can begin again, expansion of cell, development of chloroplast and the growth of roots is supported by such stored reserves.

The dispersal unit of a plant is the seed. It is capable of surviving the phase involving seed maturation to the formation of the following generation in the form of a seedling after its germination. Primarily in its dry state, the seed is well-resourced for sustaining adverse conditions for days, thus, helping in its survival.

The seed would go in a dormant state to augment germination over the course of time. The pre-harvest germination would also be prevented through dormancy. To get better knowledge of the ways to control germination through several environmental factors and applied chemicals, several studies have been performed.

Nevertheless, not much would be known regarding the process of emergence of embryo from the seed to consummate germination and the way which would obstruct embryo emergence in dormant seeds (Bewley, 1997).

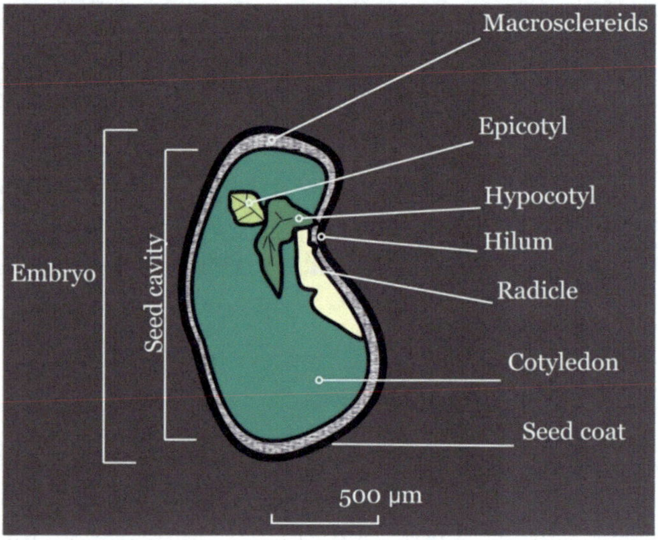

Figure 7.2: Seed and Embryo.

Source: Image by Wikimedia Commons

Environmental aspects like light, temperature and time of drying and storing seeds along with genetic factors control seed dormancy.

The application of genetics and molecular genetics in Arabidopsis elucidates few factors associated with dormancy and germination mechanisms through mutants and genes being identified, responsible for regulating such processes.

In recent times, there have been many reviews (Finch Savage and Leubner-Metzger 2006, Finkelstein et al., 2008, Holdsworth et al., 2008a) unfolding several details pertaining to the molecular mechanisms derived from molecular genetics largely in combination with physiological experiments.

Seeds form a major part of the world's diet. 90% of the total cultivated seeds is comprised of cereal grains alone. Half of the global per capita energy consumption is supposed to be contributed through them.

Hence, it would not be surprising if seed biology is considered as amongst the most widely researched fields in plant physiology. As per a casual study of the Agricola database, more than 5000 seed germination publications and 700 associated to seed dormancy had been featured by 2000.

There are few uncertainties related to two primary questions; how an embryo emerges from the seed completing the cycle of germination, and the other would be, how the embryo is obstructed to emerge so as to continue the dormant state of the seed?

Though the subject is considered to be vast and incomprehensible, yet, it would have an overview of the important processes of germination and a report on all the probable obstructions on the related subject which might result in dormancy.

The seed would lead to the following generation of plants. It is supposed to be structurally and physiologically prepared to be dispersed, besides holding the embryo in the form of a new miniature plant. It would be well nourished with food reserves for sustaining the growth of the seedling till the time it becomes an independent plant and an autotrophic organism.

Since a seed is supposed to produce a new plant, the existence of dormancy, which fundamentally obstructs germination, might seem odd. However, spontaneous germination might not be beneficial for a seed.

For example, vegetative growth following the growth of progeny would be possible only if annuals germinate in spring. While, if germination takes place at the time of fall, the cold winters could damage the vegetative plants. Thus, as an adaptive feature, dormancy would augment the distribution of germination in the tribe of seeds over time.

Figure 7.3: Dormant Seeds in Cold Winters.

Source: Image by PxHere

In agricultural crops where there would be a requirement of fast germination, seed dormancy is considered to be an unwanted trait. Nevertheless, a certain level of dormancy is beneficial regardless, at the time of seed development. In the case of cereal crops, it would be specifically correct as it would prevent grains from germinating even though it would be attached to the parent plant (preharvest sprouting). As a result, the agricultural industry would incur huge losses.

A majority of the dormancy mechanisms existing in the seeds pertaining to their wild ancestors would have been eliminated through wide-spread domestication and crop species breeding. However, dormancy might reemerge due to unfavorable environmental conditions.

On the contrary, inherent mechanisms of dormancy, wherein, seeds would continue to be embedded in the soil for years before finishing the process of germination, would often result in maturing the weed seeds.

7.2. SEED DEVELOPMENT

There would be two main stages involved in the development of a seed; seed development and seed maturation. A single cell zygote formation would start from embryogenesis, a stage of morphogenesis. It would end once all the embryo structures would be formed, that is in the heart stage (Mayer et al., 1991). Following would be the growing stage in which the seed sac would be filled by the embryo (Goldberg et al., 1994).

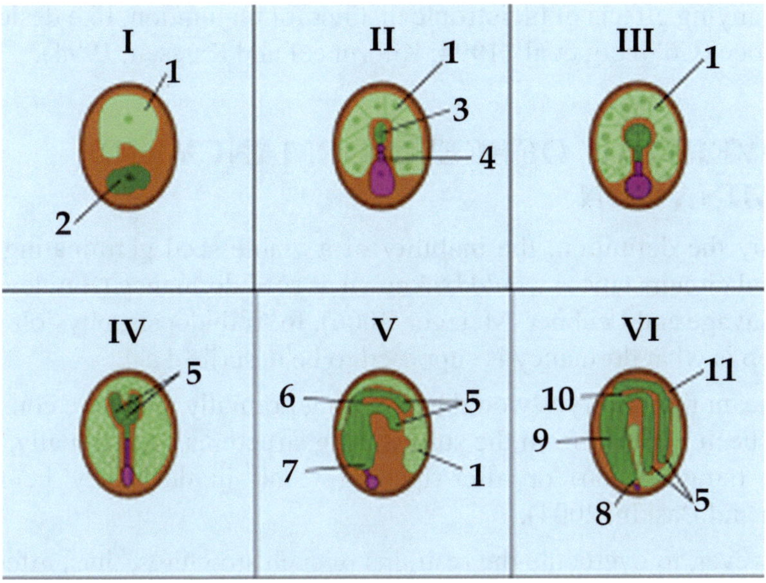

Figure 7.4: Seed Development Cycle.

Source: Image by Wikimedia Commons

Cell division in the embryo would pause by the end of the growing stage of an embryo (Raz et al., 2001). Thereafter, the seed with a well-developed embryo would experience maturation at the time when food reserves would get stored, developing tolerance for dormancy and desiccation (Goldberg et al., 1994).

There have been many comprehensive studies on the developmental (Baud et al. 2002), metabolic (Fait et al., 2006), proteomic (Fu et al., 2005; Gallardo et al., 2001, 2002; Rajjou et al., 2004 and Chibani et al., 2006) aspects and changes in expression of genes (Girke et al., 2000; Nakabayashi et al., 2005; Finch-Savage et al., 2007) at the time of seed development, pertaining to dry seeds and following germination.

Developments of seeds are considered to have been widely studied through mutants defective in multiple facets of the process. Lethality or forming a seedling phenotype would be due to the affected mutants from the morphogenesis stage (Mayer et al., 1991; Meinke, 1995).

Seed germination mutants would include characteristics related to germination and dormancy which would be affected and occasionally accompanying effects of pleiotropic distinct for maturation, like desiccation intolerance (Goldberg et al., 1994; Koornneef and Karssen, 1994).

7.3. OVERVIEW OF SEED DORMANCY AND GERMINATION

Going by the definition, the inability of a viable seed germinating under congenial circumstances would be known as seed dormancy (Bewley, 1997; Finch-Savage and Leubner–Metzger, 2006). In Arabidopsis, physiologically non-deep is what dormancy is supposed to be described as.

It meant that embryos would be growing normally that were considered to have been released from the surrounding structures. Additionally, moist-chilling (stratification) or after-ripening result in dormancy being lost (Baskin and Baskin, 2004).

However, to overcome the restrains of such structures, thus, affecting a seed's dormancy state, growth possibilities in relation to the embryo would be vital together with the surrounding layer of the testa and endosperm of the embryo (Kucera et al., 2006). Detecting definite genetic and physiological variations would be challenging due to the event of dormancy being restricted at the various stages of development and with the environmental factors.

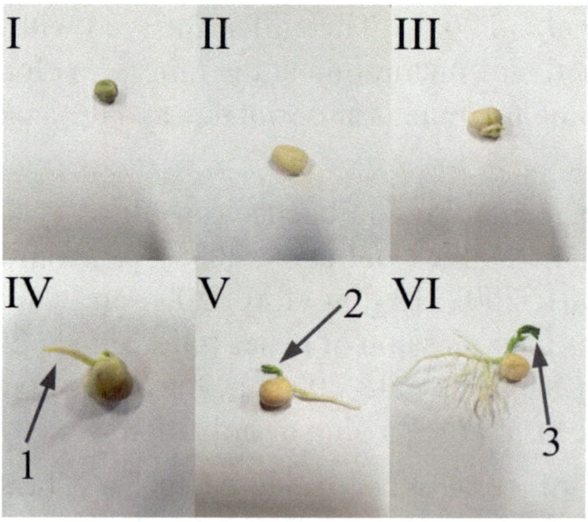

Figure 7.5: Stages of Germination.

Source: Image by Wikimedia Commons

All dormancy assays being grounded on seed germination gives rise to such challenges. It would be due to the balance between the level of dormancy and the ability of the embryo for overcoming dormancy. Automatically, factors affecting dormancy and germination can be distinguished based on their impact on germination acting as either an inhibitor or a promoter.

Advanced or rapidly germinating mutants would represent dormancy promoting genes or the ones repressing germination. Further, it can be distinguished by interpreting the time and the location of action in relation to such factors (at the time of maturation or inhibition of the seeds within the embryo, the endosperm or the in the testa).

The association of such factors, their huge environmental impact, equally at the time of development of the seed as well as at the time of inhibition makes seed dormancy as a very intricate trait.

As per the definition, the process starts with the absorption of water by the dormant dry seed and ends with the embryonic axis being elongated: germination (Bewley and Black, 1994). There would be three phases in the process of water absorption by the seed:

- Phase 1 – rapid initial absorption
- Phase 2 – plateau phase
- Phase 3 – additional increase in water absorption, but only at the time of germination (Schopfer and Plachy, 1984; Bewley, 1997; Manz et al., 2005).

The recommencement of vital processes comprising of transcription, translation and DNA repair subsequently with cell-elongation and finally recommencement of cell division during radicle protrusion would be regarded as the initial signs of germination (Barroco et al., 2005; Masubelele et al., 2005). Physically, the process of germination occurs in two stages, wherein rupturing of the testa would eventually lead to the rupturing of the endosperm.

Germination would be completed after the micropylar endosperm is ruptured by the emerging radicle (Karssen, 1976; Hepher and Roberts 1985; Leubner-Metzger et al., 1995; Krock et al., 2002; Petruzzelli et al., 2003; Leubner-Metzger, 2003; Liu et al., 2005).

In Arabidopsis, performing germination assays would frequently take place in the presence of light on the seeds that would have been harvested freshly have been accumulated for a short time (LéonKloosterziel et al., 1996a).

The frequency of germination post- different stages of cold treatment (Cutler et al., 1996) as well as the process of germination occurring in the absence of light (Meng et al., 2008; Kim et al., 2008) are additional parameters. For investigating genetic differences at the initial stages of seed development, the germination of undeveloped seeds removed either from the silique or from within the detached fruits of the plant could be examined apart from examining mature seeds (Raz et al., 2001).

Seed germination is contributed to by both maternal (testa) and zygotic origin (embryo and endosperm) tissues, hence, the origins of such tissue variations is supposed to be taken into consideration by the genetic analyses of seed dormancy.

Maternal properties in comparison to zygotic aspects are inherited maternally probably due to the genetic composition of the testa, which surround the embryo. However, differences in genes in association with factors transmitted by the mother plant into the seed, might also be one of the reasons.

The deduction of maternal inheritance made on the basis of germination of seeds can be achieved after reciprocal crosses by using parental genotypes in the form of both female and male parents. The fertilization would result in producing the endosperm.

However, in the case of triploid tissues, the female parent contributes two times more genomic content compared to the male parent. In certain situations, there might be disparities occurring in reciprocal crosses with the involvement of reciprocal crosses.

Recent studies have revealed function of the single endosperm layer during dormancy: the requirement to debilitate the particular layer of the cell for gemination (Müller et al., 2006; Bethke et al., 2007a; for review Holdsworth et al., 2008a).

In *Arabidopsis*, the factors responsible to promote germination like after-ripening, light, cold treatment (known as stratification) help to overcome seed dormancy. Additionally, applied chemicals, like gibberellin and KNO3 (Derkx and Karssen, 1993a; Alboresi et al. 2005; Ali-Rachedi et al. 2004), NO (Bethke et al., 2007b) constitute properties pertaining to the promotion of germination.

For germination, neither of the mentioned environmental aspects are a complete necessity due to the interdependency of the requirement for one factor over the other. It would be similar to the interaction between light and temperature, as depicted by Cone and Spruit (1983).

Such necessities related to exogenous factors would hugely be dependent on the genotype. Releasing dormancy through dry after-ripening, or due to cold or nitrate exposure, subsequently by the exposure of light is supposed to have been studied in the accession of the Cape Verde Islands (Cvi).

It seemed, the sensitivity of the seed towards cold, nitrate as well as light was reliant on the time duration of the seeds being dry after-ripening.

The seeds would be developing sensitivity first towards nitrate, followed by cold and eventually towards light (Finch-Savage et al., 2007). Moreover, it was revealed that the increased level of sensitivity towards environmental signals would be uncertain, with seed production being done over different years, responding differently.

Thus, being consistent, dormancy would not solely be established genetically, but also by the environmental conditions at the time seeds are formed (Donohue, 2005).

Non-dormant seeds being exposed to hostile germination conditions for a period of time (for example, soaked seeds kept at a comparatively high temperature in the absence of light or in darkness), would once again arrive at a dormancy state, known as secondary dormancy (Cone and Spruit, 1983; Derkx and Karssen 1993b; Hilhorst, 2007).

Identifying the characteristics of the vital regulator(s) that would result in preventing the start of germination (dormancy) and generating the process of germination along with their mutual interaction, would be considered as a difficulty in the field of dormancy and germination. Moreover, knowing the ways by which environmental factors, like light and cold would be inducing the endogenous factors controlling germination, is considered to be important.

7.4. CLASSIFICATION OF SEED DORMANCY

A dormancy classification system was devised by Marianna G. Nikolaeva, which reflects that dormancy would be controlled by both morphological and physiological characteristics of the seed (Nikolaeva, 1967, 2004). Moreover, an exhaustive classification system was considered to have been proposed by Baskin & Baskin (1998, 2004) on the basis of the above -mentioned system. It comprises of five groups of seed dormancy:

- Physiological (PD)
- Morphological (MD)
- Morphophysiological (MPD)
- Physical (PY)
- Combinational (PY + PD)

7.4.1. Physiological dormancy (PD)

PD is considered as the most infinite form and is supposed to be found in the gymnosperm seeds and in every main angiosperm clades. In case of temperate seed banks it would be the most common form of dormancy as well as the most infinite group of dormancy 'in the field'.

'In the laboratory', for a majority of seed model species, the main form of dormancy would be the PD and it includes *A. thaliana, Helianthus annuus, Lactuca sativa, Lycopersicon esculentum, Nicotiana spp., Avena fatua* and various cereals. Hence, in most the reviews, PD is given the prime focus. PD could be categorized into three stages: deep, intermediate and non-deep (Baskin & Baskin, 2004).

PD Deep

Embryos when removed from such seeds would either stop growing or abnormal seedlings would be produced. Their dormancy would not be disrupted by GA treatment and there would a requirement of quite a few months of cold (subtype a) or warm (subtype b) stratification prior to commencement of germination (Baskin & Baskin, 2004; Baskin *et al.*, 2005). For example, *Acer platanoides* (PD deep) and *Acer pseudoplatanus* (PD intermediate) (Aceraceae, Finch-Savage *et al.*, 1998).

PD Non-deep

A huge number of seeds show non-deep PD (Baskin & Baskin, 2004). Normal seedlings would be produced by the embryos coming out from such seeds; the dormancy could be disrupted by the GA treatment. Moreover, scarification, after-ripening in dry storage and cold and warm scarification could also disrupt dormancy, depending on the species.

On the basis of the changing patterns in physiological responses to temperature, nondeep PD could be categorized into five types. However, majority of seeds would belong to the type 1 or 2 category. In this case, the level of temperature for seeds to germinate would moderately increase at the time of development of non-deep dormancy: low to higher (type 1, e.g. *A. thaliana*) or higher to lower temperature (type 2).Additionally, with nondeep PD being moderately released, the seeds sensitivity to light and GA would increase.

7.4.2. Morphological dormancy (MD)

MD is apparent in seeds where the embryos would not be fully developed (in terms of size), or undifferentiated (e.g. into cotyledons and hypocotyl-radical).Such embryos would basically require sufficient time for growth and germination and would not be considered as (physiologically) dormant. For example, celery (*Apium graveolens*) (Jacobson & Pressman, 1979).

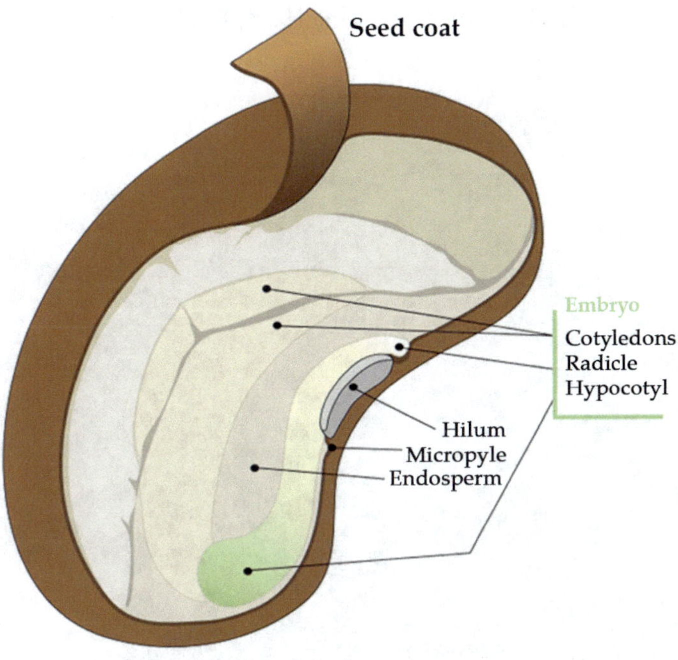

Figure7.6: Various Components of a Seed.

Source: Image by Wikimedia Commons

7.4.3. Morphophysiological dormancy (MPD)

MPD is apparent in seeds where the embryos would not have developed fully. However, their dormancy would include a physiological constituent in it (Baskin & Baskin, 2004). Thus, a dormancy-breaking treatment is something that would be needed by such seeds.

For example, a distinct combination of warm and/or cold stratification or GA application in few cases. There are eight MPD stages known. For example, *Trollius*, (Hepher & Roberts, 1985), *Fraxinus excelsior* (Finch-Savage & Clay, 1997).

7.4.4. Physical dormancy (PY)

Palisade cells within the seed or fruit skin controlling the movement of watercause PY through their water-impermeable layers. The PY dormancy could be disrupted by mechanical or chemical scarification. For example, Melilotus Trigonella (Baskin & Baskin, 1998; Baskin, 2003)

Figure 7.7: Seeds with Physical Dormancy.

Source: Image by Wikimedia Commons

7.4.5. Combinational dormancy (PY + PD)

Seeds constituting water-impermeable coats (like in PY) with a combination of physiological embryo dormancy would cause PY + PD apparently (Baskin & Baskin, 2004). For example, Geranium and Trifolium.

7.5. REGULATION OF DORMANCY BY HORMONES AND OTHER SMALL MOLECULES

7.5.1. Dormancy Induction and Maintenance

Substances which would accumulate in seeds correlating to the dormancy depth and genetically analysing the lines as per various degrees of dormancy are identified as potential dormancy regulators.

Abscisic Acid

There have been various lines of evidences, which determine dormancy being induced by abscisic acid (ABA) at the time of embryo maturation (reviewed in Feurtado et al., 2007; Finch-Savage et al., 2006; Finkelstein et al., 2002; Kermode, 2005; Nambara et al., 2003). Germination can be inhibited by ABA correlating the accumulation with the beginning of dormancy.

ABA -deficient seeds lower dormancy as a result of mutations, chemical inhibition of biosynthesis, or isolating of ABA by antibodies expressed in seeds. On the contrary, enhanced dormancy would be a result of an overexpression of ABA biosynthesis enzymes (Lin et al., 2007;Nambara, et al., 2003).

As per genetic studies, maternal tissues producing ABA or an exogenous supply of ABA is enough for inducing dormancy (Finch-Savage et al., 2006; Finkelstein et al., 2002,). In other words, the form of dormancy being controlled by the embryo is reliant on ABA synthesis within the embryo or endosperm.

Although in dry dormant seeds, ABA declines comparatively to lower levels, due to distinct ABA 8 – hydroxylases catabolism, encoding the P450 CYP707A cytochrome family (Okamoto et al., 2006), dormancy is continued by renewing the ABA accumulation resulting in imbibition of dormant seeds (Ali-Rachedi et al., 2004;Grappin et al., 2000;Le Page-Degivry et al., 1992).

Members related to the AtNCED (9-cis-epoxycarotenoid dioxygenase) family :NCED6 and NCED9)-, vital for synthesizing ABA for both embryo and endosperm at the time of inducing dormancy, are identified by expressing and genetic analyzing particular genes that encode main enzymes ABA metabolism (Lefebvre et al., 2006).

Identification of particular genes associated with maintaining dormancy [NCED6, NCED9, and ZEP (zeaxanthin epoxidase)] as well as releasing dormancy through ABA catabolism (CYP707A2) have been attributed by means of further studies (Cadman et al., 2006; Millar et al., 2006).Therefore, the enzymes expressed differentially by means of combined action ,as well as various steps of synthesis and catabolism control the ABA levels in seeds resulting in dormancy.

Dehydration and Desiccation Tolerance

The huge majority of dormant seeds are considered to be "orthodox". The desiccation tolerance by ABA would be induced late in the development of the seed, and allowing them to survive in the dry state for a longer time. On the contrary, "seeds, which are "recalcitrant" would not be considered tolerant towards desiccation.

Moreover, it would be associated with active oxygen species (AOS) generation as well as oxidative damage at the time of dehydration (Bailly C, 2004; Berjak P., 2006). As a result, it would be suggesting that the dependency of desiccation tolerance would be in part related to the scavenging ability of AOS compounds through antioxidant defense systems.

The catalase activity is supposed to have been increased by most maturing seeds (Bailly C., 2004;Berjak P., 2006) along with stored glutathione and peroxiredoxins (Haslekas et al., 1992; Stacy et al., 1996) or even storing proteins which would serve as substrates for the purpose of oxidative carbonylation (Bailly C., Job et al., 2005). Although, for maturation, the production of glutathione in the embryo would seem to be important (Cairns et al., 2006)and hypersensitivity would be conferred for numerous germination-inhibiting stresses by overexpressed peroxiredoxins, yet, there is association between the peroxiredoxin levels and the time of dormancy (Haslekas et al., 2003).

The desiccation tolerance would incorporate forming an intracellular "glassy state" caused by a combination of sugars being supersaturated with late embryogenesis abundant (LEA) proteins (reviewed in Berjak P., 2006). The levels of LEA transcript do seem to be associated with dormancy even though there the LEA proteins and desiccation toleration are correlated (Baumbusch et al., 2004;Finch-Savage et al., 2007).

Seed Coat Components

Altered dormancy and longevity of seeds belonging to various species would be a result of genetic difference pertaining to the structure and pigmentation of the seed coat (testa) or the layers surrounding it, like the pericarp of grains (Debeaujon et al., 2007; Groos et al., 2002; Sweeney et al., 2006). The testa provides protection to the embryo both from damages caused by the environment as well as restricting the growth process before germination.

Pigmentation affected by mutation is inclusive of pigment biosynthetic genes being disrupted, involving transporters of the pigment compartmentation and regulating the production of the biosynthetic enzymes through transcription factors (reviewed in 106).The pigments in seed coats are normally phenolic compounds, like flavonoids. Depending on the species the pigments are accumulated; grains commonly have red phlobaphene, legumes have is flavones and Arabidopsis has proanthocyanidins as well as flavonol glycosides (reviewed in Lepiniec et al., 2006).

Polymers usually have the proanthocyanidins, which are be cross-link through oxidation to wall elements during seed maturation. It results in the cell layers being more thick providing superior mechanical restraint along with the permeability reducing for water, gases and hormones.

Additionally, flavonoids restrict the availability of oxygen further as an effective antioxidant, thus, creating room for metabolic processes inhibition relevant to after-ripening and germination, e.g., ABA oxidative degradation. Seeds pigments are regulated by light senses.

The pigmentation degree is dependent on experiencing photoperiod by the maternal plant at the time of seed setting along with the filtered light pertinent to seeds that are extremely pigmented and reaching the embryo. Though it has been revealed that the flavonoids would absorb only the UV light (Winkel-Shirley B., 2002).

Flavonoids are considered to be protective against antimicrobial and antiherbivory functions, UV radiations as well as perhaps oxidative damage. Therefore, breeding programs attempted for the isolating characters with regards to the color of the seed, dormancy and longevity would suggest the features to be inseparable.

For example, recording few loci that affect the color of the seed, dormancy and few stress features of rice would reveal very close connection so as to refer the traits as domestication block; a single locus might be controlling few of such features (Ji HS et al., 2006).

Nevertheless, while pigmented seeds would be considered to be usually more dormant, the non-pigmented seeds dormancy can be augmented through added features like hormone levels or sensitivity (Flintham JE., 2000;Gale et al., 2002;Walker-Simmons M., 1987).

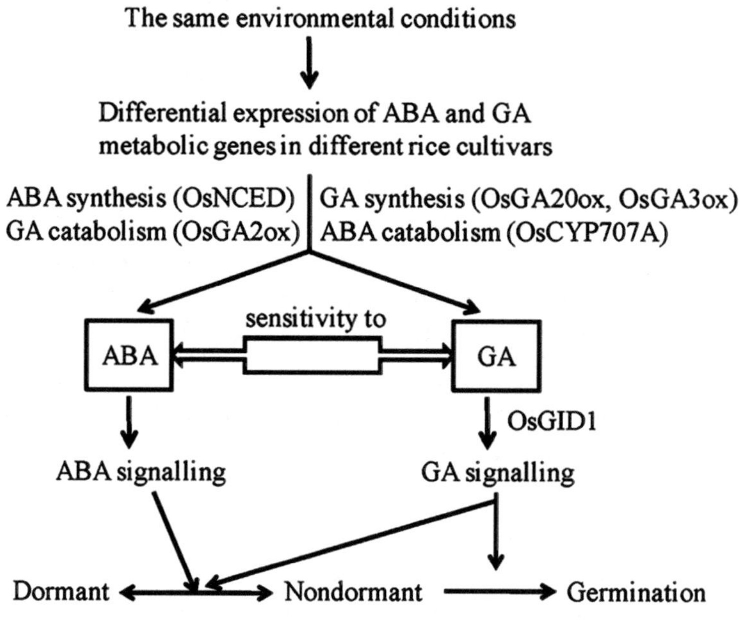

Figure 7.8: Seed Dormancy in Rice.

Source: Image by Wikimedia Commons

7.5.2. Dormancy Release and Germination

Different mechanisms help in achieving dormancy release. It constitutes intricate interactions in association with the environment, facilitated by phytohormones as well as additional small molecules.

Moreover, the conditions selected for germination would the ones that are most favourable for the survival of the plant. Following are the molecular mechanisms of dormancy release by means of gibberellins, brassinosteroids, ethylene, reactive oxygen species and nitrogenic compounds, like nitrate as well as nitric oxide (NO).

Gibberellins

Gibberellins [e.g., gibberellic acid (GA)] belong to a family comprising of 136 tetracyclic diterpenes. Being active in the form of plant hormones, a small subset of it can stimulate the germination of seed in a variety of plant species. The species would determine the predominance of the active GA (reviewed in Thomas et al., 2005).

Germination is stimulated by the gibberellins by means of inducing hydrolytic enzymes, making the blocking tissues weak, like the endosperm or seed coat to induce mobility within the storage reserves of the seed and incite development of the embryo (reviewed in Bewley et al., 1994).

Germination might even be incited by GA through an embryonic to vegetative development transition, partially facilitated by PICKLE (PKL), the chromatin remodeling factor (Henderson et al., 2004).

In the transition, the stimulation of GA is recommended by augmenting the PKL embryonic root phenotype by the inhibitors of biosynthesis of GA and the disappearing of the embryonic identity protein FUSCA 3 (FUS3) by means of GA stimulation. It controls ABA synthesis positively and the GA synthesis negatively (Gazzarrini et al., 2004; Henderson et al., 2004).

In dormancy release, the function of gibberellins is considered to be controversial. Although there is an association between GA accumulation, dormancy release and germination, yet, germination pertaining to all species or in completely dormant Arabidopsis seeds isn't stimulated by GA treatment alone (Ali-Rachedi et al., 2004; Bewley JD., 1997; Derkx et al., 1994). Prior to GA levels and sensitivity would be increased, a lowering of ABA levels might be required (Ali-Rachedi et al., 2004, Jacobsen et al., 2002).

Nevertheless, both GA and light sensitivity increase with the development of after-ripening in Arabidopsis (Derkx et al., 1993). Moreover, in Arabidopsis, the dormancy- relieving effect of moist chilling is facilitated by the increasing of GA levels (Yamauchi et al., 2004).

GA synthesis is promoted by light and ABA degradation in imbibing lettuce seeds is promoted by light and GA (Toyomasu et al., 1998).

An increase of 40 fold in after-ripened seeds occurred by the accumulation of GA3ox2 (GIBBERELLIN 3 OXIDASE), the GA biosynthetic gene transcript. The expression of GA2oxl (GIBBERELLIN 2 OXIDASE) is at the highest levels in GA deactivating enzyme in extremely dormant seeds of Arabidopsis ecotype Cvi (Cape Verde Islands) (Finch-Savage et al., 2007).

The expression of genes in biosynthesis- GA20oxl (GIBBERELLIN 20 OXIDASE), GA20ox2 and GA3oxl increased as a result of stratification while the GA catabolic gene GA2ox2 decreased in expression (Yamauchi et al., 2004). Altogether, the start of after-ripening might not be stimulated by GA, but it might be considered essential, however, not enough for dormancy release in seed and germination.

In dicots such as Arabidopsis and tomato, GA is considered important for the purpose of seed germination as mutants with defective genes fail to germinate while encoding GA biosynthesis enzymes (Mitchum et al., 2006; Steber C., 2007 and references within).

Exogenous application of GA, ABA biosynthesis or signaling affected by mutations or the mechanical restraint of the testa or endosperm removal result in relieving the non-germination of GA mutants. Therefore, GA mutants matching the standard definition of dormant, would recommend GA synthesis as one of the necessary steps required for releasing dormancy.

Brassinosteriods

Brassinosteroids (BRs) considered as plant steroid hormones are associated with the elongation of the stem and unfurling of the leaf (reviewed in Clouse S., 2001) promoting seed germination at the same time. The BR biosynthesis and signaling mutants are extra- sensitive compared to wild-type towards ABA inhibition of seed germination, even though they germinate effectively. Hence, it signifies that the potential to germinate is decreased by such mutations (Clouse et al, 1998; Steber et al., 2001).

To overcome nongermination of GA biosynthesis and sleepy 1 (sly1) GA signaling mutants in Arabidopsis, the application of epibrassinolide (EBR) and brassinolide (BL) is required; though different mechanisms are involved for stimulating germination in tobacco seed (Leubner-Metzger G., 2001).

While the rupturing of endosperm in non-photo dormant tobacco seeds imbibed in the dark would be stimulated by both BR and GA, the class I β-1,3 glucanase (bGLU I) activity, crucial for rupturing endosperm of photo dormant tobacco seed, would be induced by GA solely (Leubner-Metzger G., 2001).

Ethylene

Breaking dormancy by ABA signaling interactions is promoted by ethylene. Seeds with ethylene resistant (etr) 1 receptor mutants and ethylene insensitive (ein) 2/ enhanced response to aba (era) 3 mutants show high dormancy associated with high sensitivity in relation to ABA in germination of seed as well as increasing of ABA synthesis (Beaudoin et al., 2000;Chiwocha et al., 2005;Ghassemian et al., 2000).

Whereas, constitutive triple response (ctr) 1 mutation and treating wild-type seeds with ethylene precursor 1- amino cyclopropane-1 carboxylic acid (ACC) results in lowering the ABA sensitivity (Ghassemian et al., 2000). Over- accumulation of GA occurs in the etrl-2 mutant, signifying that over accumulation of ABA leading to increased dormancy is compensated for to a certain extent by higher GA accumulation. Thus, seed germination by ethylene stimulation occurrence would take place by antagonizing ABA signaling.

Nitric Oxide, Nitrate, and Active Oxygen Species

Several compounds containing nitrogen: NO gas, nitrite (NO2 -), nitrate (NO3 -), nitrogen dioxide, ammonium, azide and cyanide are involved in promoting dormancy release as well as germination of seed in several species (reviewed in Bethke et al., 2007).

Even at the time of development of the seed, the nitrate provided by the maternal plant results in lowered dormancy (Alboresi et al., 2005). The analysis of their mechanism (s) is complicated by the compounds being interconverted and NO donors, presence of scavengers and detection equipment lacking specificity. However, they might be acting as NO sources or through similar NO mechanisms (Bethke et al., 2007).

Nitrate reductase (NR) derived from nitrite and nitrate (reviewed in Bethke et al., 2007) and probably nitric oxide synthase (Crawford et al., 2006), would primarily result in the production of enzymatic NO in plants.

But, even though there is a correlation between Arabidopsis NRI expression and dormancy breaking conditions (Finch-Savage et al., 2007), nitrate is accumulated and dormancy reduced by seeds of NR -deficient plants (Alboresi et al., 2005).

The oxidizing environments of glyoxysomes involving lipid catabolismalso result in the production of NO (Corpas et al., 2001). The apoplast or soil can converting nitrite into NO with low pH (Bethke et al., 2004).

Such compounds with nitrogen lead to promoting seed germination by metabolism effects, oxidation state or signaling (Bethke et al., 2007). The expression of few gene encoded enzymes in the pentose phosphate pathway are induced by nitrate for dormancy relieving conditions, providing vital nutrients for germination (Finch-Savage et al., 2007).

The pentose phosphate pathway is stimulated by NO, nitrite as well as nitrate. Hence, it leads to germination through increased oxidation of NADPH to NADP+ also as the limiting electron acceptor. Increasing accumulation of H2O2 from the β-oxidation of fatty acids stored in seeds might occur as a result of the NO inhibition of catalase.

Germination is stimulated by accumulated H2O2, which would act as a substrate for peroxidases to lead to NADPH to NADP+ oxidation or ABA breakdown (reviewed in Bailly C., 2004). Hormone levels canalso be modified by nitrate by inducing enzyme expression catalyzing deactivation of ABA (CYP707A2) and GA biosynthesis (GA3oxl) (Finch-Savage et al., 2007). Vacuolation and weakening of the cell wall of the aleurone layer would occur if germination may well be stimulated by NO in Arabidopsis (Bethke et al., 2007).

7.6. GENETIC VARIATION FOR GERMINATION CHARACTERS

Genetic differences can be induced by mutagenesis, however, would be existing even among several natural accessions (ecotype) based on Arabidopsis. Genetic differences in relation to germination can be noticed while comparing genotypes amid identical environments. It would be inferred that apart from the conditions of germination, the growth conditions at the time of development of the seed and conditions related to storage (including the storage time of the seeds) is supposed to be same. Additionally, discriminatory test conditions would be required between genotypes.

A huge number of assays would result in germination tests being used proficiently in the case of mutant screens. However, genetic misclassification pertaining to individual seeds can be caused by germination trait differences hence, resulting in false-positives in mutant screens.

A low level of dormancy is shown in the Arabidopsis genotypes: *Landsberg erecta* (Ler) and Colombia (Col), and majorly being used in Arabidopsis research.

After almost one month of after-ripening, dormancy disappears (van der Schaar et al., 1997). Such mutations in dormancy genes would not be possible. However, by using extra dormant accessions, the problem can be dealt with (Koornneef et al., 2000).

7.6.1 The Natural Variation Analysis

Arabidopsis being an annual plant, huge variations in dormancy can be seen in nature (Ratcliffe, 1976). The first genetic study on the variations of germination between Arabidopsis accessions was performed by Kugler (1951). As depicted by Quantitative Trait Loci (QTL) analysis, the dormancy variations are amenable to genetic analysis in Arabidopsis, intended for germination of seed or dormancy features in many recombinant inbred line (RIL) populations (van der Schaar et al., 1997; Alonso Blanco et al., 2003 and Clerkx et al., 2004).

Such RIL populations are supposed to have been created through cross accessions with various levels of seed dormancy in terms of standard laboratory accession Ler. The Ler/Col RILs are grown in three independent greenhouse experiments and were tested in various germination conditions. Overall, the identification of 14 QTL was done wherein, for majority the effects were minor, in spite of few dormancy variations existing between the parental lines.

Several of them are located only in particular conditions of germination or in seeds after one or two dates of harvest. It would signify the intricacy of the genetic control of dormancy variations. Moreover, seed germination analysis responding towards cold and dark displayed a fine mapping possibility with QTL cloning accountable for cold-tolerant, dark germination (Meng et al., 2008).

7.7. THE GENETIC CONTROL OF SEED STORAGE COMPOUNDS, SEED DETERIORATION AND DEVELOPMENT OF EARLY SEEDLING

The survival of seeds can extend for a long time without germination. It can be possible either while being stored in dry conditions by getting buried in the soil. Such seeds create the seed bank that wait till the time the environmental conditions would be appropriated for germination. In spite of tolerating extreme desiccation, several seeds, even under dry conditions storage would ultimately result in a viable loss.

As per several suggestions, compounds like few sugar and proteins, for example, LEA (Late embryogenesis abundant) (Skriver and Mundy, 1990) as well as Heat shock proteins (Hsps) (Wehmeyer and Vierling, 2000; Hong and Vierling, 2001), accumulating at the later seed development levels play a protective role in desiccation.

It is crucial for a seed which would have survived a no germination period, to be able to grow into strong seedling so as to compete with other seedlings. The availability and mobilization of stored resources probably play a major role in formation of the seedling.

The storage proteins and lipids accumulation are considered to be faulty seed maturation mutants, like lec1, lec2 and fus3, wherein, even if they would be germinating on filter paper, however, it would not be possible to transfer them competently from the substrate to soil compared to other seeds. Moreover, to form a seedling, sugar supplemented media are required prior to being transplanted to the soil. The storage compounds of seeds with lowered seed oil content are be particularly be affected by the wrinkled1 (Focks and Benning, 1998) mutant.

Additionally, it depicts a reduction in the prospect of germination reducing the formation of seedlings and responding differently to sugar as well as ABA signals compared to the wild type (Cernac et al., 2006). An AP2/EREB family transcription factor is encoded by WRI1 (Cernac and Benning, 2004), with LEC2 regulating it (Baud et al., 2007). The carbon metabolism pertaining to developing seeds is controlled by WRI1 to act for the accumulation of oil in seeds.

At the time of germination and growth of early seedlings, the mobilization of food reserves needs lipid activity and degrading enzymes of carbohydrates. Breaking down of the stored lipid is regarded as a complex activity with several steps. It involved the identification of several genes and their mutation effects in mutants and double mutants (reviewed in Penfield et al., 2007; Graham et al., 2008).

Fatty acids are stored in the form of triacylglycerol (TAG) in oil bodies, being hydrolyzed by lipases. Through an ABC transporter with the CTS (comatose) protein encoding, the fatty acids released are transferred to glyoxysomes (peroxisomes in which the glyoxylate cycle occur) (Russell et al., 2000; Footitt et al., 2002). It is the place for β-oxidation = with fatty acids being converted to acyl-CoA esters from acetyl Co-A and eventually getting converted to four carbon sugars.

Next the sugars are transmitted to the mitochondria converted to malate and being transmitted to the cytosol for gluconeogenesis or respiration. Absence of flavonoids from mutant seeds cts and kat2-1 have been revealed by gene expression, however in the presence of sucrose results in accumulation. Hence, it signifyies an important role of β-oxidation for flavonoid biosynthesis gene induction.

The early seedling formation is arrested by sugars. As being depicted, the process has a close association with plant hormone biosynthesis and signaling, specifically corresponding to ABA (for review see Finkelstein and Gibson, 2001; Gazzarini and McCourt, 2001; León and Sheen, 2003; Rook et al., 2006; Dekkers and Smeekens, 2007; Rognoni et al., 2007).

Mutants considered insensitive towards the effects of glucose and sucrose inhibition in the early development of the seedling are separated by multiple groups and recognized in the form of ABA- deficient mutants (i.e. aba2/isi4/gin1/sis4 and aba3/gin5) as well as ABA insensitive 4 (abi4/sun6/isi3/gin6/sis5) or sugar insensitive (Arenas-Huertero et al., 2000; Huijser et al., 2000; Laby et al., 2000; Rook et al., 2001).

An AP2 domain with transcription factor binding a CE1-like component in various ABA and sugar regulated promoters (Finkelstein et al., 1998; Niu et al., 2002; Acevedo-Hernandez et al., 2005), is encoded by ABI4 as a matter of fact, linking sugar regulation to ABA signaling. Nevertheless, multiple other examples related to processes and genes being regulated by ABA and sugar exist.

ABI3 as a chief regulator of glucose insensitive (GIN) signaling, was revealed by Dekkers et al. (2008). It is considered to be in sync with prior reports which revealed an overexpression of ABI3 (either AtABI3 or CnABI3) in Arabidopsis with a glucose oversensitive phenotype (Finkelstein et al., 2002; Zeng and Kermode, 2004).

Further, ABA and glucose signaling are considered to be overlapping to a greater range (Dekkers et al., 2008). In recent times, DOG1 involvement and the mediation of ABA- sugar signaling pathway in seedlings have been identified. The role is distinct for the GSQ5/DOG1 Cvi allele considering addition of glucose induce the expression of the GSQ5/DOG1 Cvi while the there is no response from the Ler and Col allele (Teng et al., 2008).

7.8. "OMIC" ANALYSES RELATED TO DORMANCY AND GERMINATION

In Arabidopsis, the expression of genome at both the levels of transcriptome and proteome has been analyzed in various recent studies (reviewed by Holdsworth et al., 2008b). A key role intended for translation in germination and dormancy release has been identified. According to the reports of Rajjou et al. (2004), transcription is considered as a step meant for restricting the germination followed by imbibition.

Existence of surplus amanitin (a specific inhibitor of DNA dependent RNA polymerase II) during imbibition results in the failure of inhibiting completion of germination. On the contrary, germination is completely inhibited by a cycloheximide translation inhibitor. Thus, it indicates the value of mRNA for the purpose of translation in imbibed seeds (Rajjou et al., 2004).

Figure 7.9: Dry Seeds.

Source: Image by Fickr

Various studies in recent times have revealed the key transcriptional and protein changes occurring in the 'dry' seeds at the time of storage and such changes might be aimed towards dormancy release. Such studies act as an indication to the possibility of both accumulation of particular proteins

(Chibani et al., 2006) as well as gene transcripts (Bove et al., 2005, Leubner-Metzger, 2005).

As per a clear description provided by Holdsworth et al. (2008b), mature seeds with 5-10% water dependent on speciesare called the 'dry' state. The remaining water is not evenly distributed within the tissues of the seed and is considered to be an indication of some areas containing sufficient water for supporting gene expression (Leubner-Metzger, 2005).

7.9. CONCLUSION

Dormancy and germination are regarded complicated features, controlled by numerous genes. In the recent years, there has been a rapid increase in the number of mutants and genes. The vital role of the hormones GA and ABA from plants have been confirmed by such studies. Nevertheless, they even depicted the participation and intricate interaction with hormones of other plants, like ethylene, brassinosteroids as well as auxin.

By using transcriptomic and proteomic analyses, differences and similarities related to pathways being affected by dormancy and germination, are considered to have been refined providing a foundation.

Further studies even can include the temporal and special elements along with distinct roles and sensing mechanisms of factors relating to the environment at the time of development, storage and germination of the seed. Altogether, it would provide an in- depth understanding of such an important and fascinating biological system.

REFERENCE

1. Bentsink, L. and Koornneef, M., 2008. Seed Dormancy and Germination. The Arabidopsis Book, [online] 6, p.e0119. Available at: <https://www.ncbi.nlm.nih.gov/pmc/articles/PMC3243337/> [Accessed 21 May 2020].
2. Bentsink, L. and Koornneef, M., 2008. Seed Dormancy and Germination. The Arabidopsis Book, [online] 6, p.4. Available at: <https://www.researchgate.net/publication/221803941_Seed_Dormancy_and_Germination> [Accessed 21 May 2020].
3. Bentsink, L. and Koornneef, M., 2008. Seed Dormancy and Germination. The Arabidopsis Book, [online] 6, p.2. Available at: <https://bioone.org/journals/the-arabidopsis-book/volume-2008/issue-6/tab.0119/Seed-Dormancy-and-Germination/10.1199/tab.0119.full> [Accessed 21 May 2020].
4. Bewley, J., 1997. Seed Germination And Dormancy. [ebook] p.1. Available at: <http://www.plantcell.org/content/plantcell/9/7/1055.full.pdf> [Accessed 21 May 2020].
5. Finch-Savage, W. and Leubner-Metzger, G., 2006. Seed dormancy and the control of germination. New Phytologist, [online] 171(3), p.2. Available at: <https://nph.onlinelibrary.wiley.com/doi/10.1111/j.1469-8137.2006.01787.x> [Accessed 21 May 2020].
6. Finkelstein, R., Reeves, W., Ariizumi, T. and Steber, C., 2008. Molecular Aspects Of Seed Dormancy. [ebook] p.5. Available at: <https://naldc.nal.usda.gov/download/22193/PDF> [Accessed 21 May 2020].
7. Nonogaki, H., 2014. Seed dormancy and germinationâ€”emerging mechanisms and new hypotheses. Frontiers in Plant Sciencc, [onlinc] 5, p.1. Available at: <https://www.frontiersin.org/articles/10.3389/fpls.2014.00233/full> [Accessed 21 May 2020].
8. Penfield, S., 2017. Seed dormancy and germination. Current Biology, [online] 27(17), pp.R874-R878. Available at: <https://www.sciencedirect.com/science/article/pii/S0960982217306164> [Accessed 21 May 2020].
9. Aceraceae, Finch□Savage et al., 1998
10. Alboresi et al. 2005
11. Ali-Rachedi et al. 2004
12. Alonso Blanco et al., 2003

13. Arenas-Huertero et al., 2000
14. Bailly C, 2004
15. Barroco et al., 2005
16. Baskin and Baskin, 2004
17. Baud et al., 2007
18. Baumbusch et al., 2004
19. Beaudoin et al., 2000
20. Berjak P., 2006
21. Bethke et al., 2007
22. Bewley and Black, 1994
23. Bewley, 1997
24. Bove et al., 2005
25. Cadman et al., 2006
26. Cairns et al., 2006)
27. Cernac and Benning, 2004
28. Cernac et al., 2006
29. Chibani et al., 2006
30. Chiwocha et al., 2005
31. Clerkx et al., 2004
32. Clouse S., 2001
33. Cone and Spruit, 1983
34. Crawford et al., 2006
35. Cutler et al., 1996
36. Debeaujon et al., 2007
37. Dekkers and Smeekens, 2007
38. Dekkers et al., 2008
39. Derkx and Karssen 1993
40. Derkx et al., 1993
41. Donohue, 2005
42. Fait et al., 2006
43. Feurtado et al., 2007
44. Finch Savage and Leubner-Metzger 2006
45. Finch□Savage & Clay, 1997

46. Finch-Savage and Leubner–Metzger, 2006
47. Finch-Savage et al., 2007
48. Finkelstein and Gibson, 2001
49. Finkelstein et al., 2008
50. Flintham JE., 2000
51. Focks and Benning, 1998
52. Footitt et al., 2002
53. Fu et al., 2005
54. Gale et al., 2002
55. Gallardo et al., 2001, 2002
56. Gazzarini and McCourt, 2001
57. Gazzarrini et al., 2004
58. Ghassemian et al., 2000
59. Girke et al., 2000
60. Goda et al., 2002
61. Goldberg et al., 1994
62. Graham et al., 2008
63. Grappin et al., 2000
64. Groos et al., 2002
65. Haslekas et al., 1992
66. Henderson et al., 2004
67. Hepher & Roberts, 1985
68. Hilhorst, 2007
69. Holdsworth et al., 2008a
70. Hong and Vierling, 2001
71. Huijser et al., 2000
72. Jacobson & Pressman, 1979
73. Ji HS et al., 2006
74. Karssen, 1976
75. Kermode, 2005
76. Kim et al., 2008
77. Koornneef and Karssen, 1994
78. Krock et al., 2002

79. Kucera et al., 2006
80. Laby et al., 2000
81. Le Page-Degivry et al., 1992
82. Lefebvre et al., 2006
83. León and Sheen, 2003
84. LéonKloosterziel et al., 1996
85. Lepiniec et al., 2006
86. Leubner-Metzger, 2005
87. Liu et al., 2005
88. Manz et al., 2005
89. Masubelele et al., 2005
90. Mayer et al., 1991
91. Meinke, 1995
92. Meng et al., 2008
93. Mitchum et al., 2006
94. Müller et al., 2006
95. Nakabayashi et al., 2005
96. Nambara et al., 2003
97. Nikolaeva, 1967
98. Okamoto et al., 2006
99. Penfield et al., 2007
100. Petruzzelli et al., 2003
101. Rajjou et al., 2004
102. Ratcliffe, 1976
103. Rognoni et al., 2007
104. Rook et al., 2006
105. Russell et al., 2000
106. Schopfer and Plachy, 1984
107. Skriver and Mundy, 1990
108. Stacy et al., 1996
109. Steber C., 2007
110. Sweeney et al., 2006
111. Teng et al., 2008

112. Thomas et al., 2005
113. Toyomasu et al., 1998
114. van der Schaar et al., 1997
115. Walker-Simmons M., 1987
116. Wehmeyer and Vierling, 2000
117. Winkel-Shirley B., 2002
118. Yamauchi et al., 2004

Chapter 8

Zygotic Embryogenesis

CONTENTS

The seed with an embryo inside it produces a new plant and is supposed to be structurally and physiologically prepared for its dispersal. It has abundant food reserves for sustaining the growing seedling till the time it forms into an independent autotrophic organism. Embryo development in higher plants involve an elaborate sequence of events provisioning division of three intersecting phases. The chapter gives a detailed analysis of the development of the early embryo, including the different ways of embryo formation and the body plan (shoot root) of the embryo.

The chapter describes the initial phases of histo-differentiation, wherein, broad miotic division would be experienced and the subsequent cells differentiate to establish their primary body plan. Thenceforth, extensive maturation follows without any more cell divisions through cell expansion and accumulating reserves in the storage tissues.

Further, it highlights the termination of maturation, by a certain level of desiccation, resulting in steady metabolism reduction since water is lost from seed tissues for passage of the embryo into a presumably metabolically dormant, or inert state.

8.1. INTRODUCTION

The first cell belonging to the sporophytic generation and a product of fertilization is the zygote. It is formed as a result of female (egg cell) and the male (sperm) gamete fusion. The zygote begins its fundamental development (embryogenesis) soon after fertilization. Further, results in usual germination, if fertilization would continue without any external or internal factors disturbing it, followed by post-germination developments as well.

The zygotic embryogenesis pattern, when subjected to ideal conditions, result in the development of the morphological basis meant for comparing with other systems of embryo formation detected *in vitro* or *in vivo*. Somatic, pollen and adventitious embryogenesis along with experimental parthenogenetic development of embryo are among few of the alternate embryo formation types.

The genetically regulated embryogenic process is considered to be significant in the comparative embryology field. Few of the embryological characters, like embryogeny types, shape and structure pertaining to the mature embryo, suspensor and haustoriaare regarded as vital in developing phylogenetic relationships between taxa and advancing the classification of natural plants.

There is barely any documentation available for the embryogenesis process at structural, ultrastructural, biochemical and molecular levels. Little data is supposed to have been gathered by means of embryogenic studies relating to hybrids and mutants. Nevertheless, the taxa count, also including the ones that had been examined through classic methods, is considered to be low. Moreover, a greater count of taxa related studies would be expected to being used for verifying many of the existing generalizations, through conventional as well as modern methods. Embryogenesis studies are significant in terms of embryology, growth and development along with plant genetics. An increasing interest is displayed in the particular field of angiosperm embryology. Unraveling the embryogenesis mechanisms and functions that enable a plant to stipulate its body plan and variation in tissue patterns, is considered to be a huge problem in research.

Even though there is supposed to have been remarkable development with diverse animal systems in the particular respect (E. H. Davidson, 1993),an elaborate knowledge of the events governing the formation of plant embryo, yet, require to be comprehended. The embryo location inside the plant and its corresponding disconnection with experimental manipulation, specifically at the initial stages of embryogenesisis a hindrance while trying to achieve the goal.

The reproductive processes occur inside the floral organs, in case of flowering plants (P. H. Raven et al., 1992). The ovule, is a multicellular structure embedded under numerous layers of cell belonging to the pistil, the female organ for reproduction (Esau, 1997, ; Raven et al., 1992, Reiser et al., 1993, Gasser et al.).

It contains the egg cell of the plant. Dissection of the chief events occurring at the initial stages of higher plant development would be difficult since the formation, fertilization and embryogenesis of egg cell would take place inside the pistil.

For the purpose of investigating early events of embryogenesis in plants, isolating plant cells and fertilizing them in vitro is considered to have become suitable in the recent times (Kranz et al., 1994). Additionally, the usage of genetic approaches for identifying genes is necessary for several embryogenic processes, pattern formation being one of the constituents (Meinke, et al., 1991, Sheridan et al., 1988).

Innumerable mutants have been identified by means of genetic manipulation of Arabidopsis thaliana by chemical mutagenesis (Meinke et al., 1985, Amasino et al., 1993, Mayer et al. 1991, Mayer et al., 1993, Berlath et al., 1993) as well as insertional mutagenesis (Forsthoefel et al., 1992, Castle et al., 1993, Shevell et al., 1994).

The post-embryonic adult phase in the life cycle of a plant includes the occurrence of major variations in events in case of higher plants. Therefore, plant embryogenesis is chiefly concerned with the basic shoot-root body pattern formation of the plant and food reserve accumulation being used by the germinating seedling after the embryonic dormancy period, inside the seed.

Identification of genes providing new understanding related to the formation of embryos at the time of plant development has been achieved in recent studies of genetics in *Arabidopsis*. The elementary processes controlling plant embryogenesis have been revealed by such studies and others with the help of molecular approaches.

In higher plants, the developmental stage of the zygote experiencing a sequence differentiating events and resulting in the creation of a mature embryois explained through zygotic embryogenesis.

Development of vital organs of the embryo along with shoot root apical meristems is occurring despite partitioning events alongside the apical-basal axis and several such events are directed by the hormone auxin.

The three embryonic tissue systems are occurring in parallel to a perpendicular radial axis with the apical-basal axis. The zygotic embryo, on maturing, is developmentally confined, metabolically inactive and being embedded inside the maternal tissues pertaining to the seed.

8.2. THE EMBRYO

The foremost and possibly most persistent query would be understanding a plant embryo. Going by the general definition, an embryo is would be regarded as a new organism which is unborn or unhatched during its initial developmental stages. Apparently, it is a product of fertilization in organisms reproducing sexually.

Currently, there areproblems faced by many in defining a plant embryo. A zygote is formed by conjugation of green algae belonging to the family of Charophycea: known to be sisters of land plants.

It would occur in such a manner that after meiosis, it would give rise to spores. Acknowledging such a type of embryogenesis would create room for reconsidering 'embryophytes' as the name for land plants.

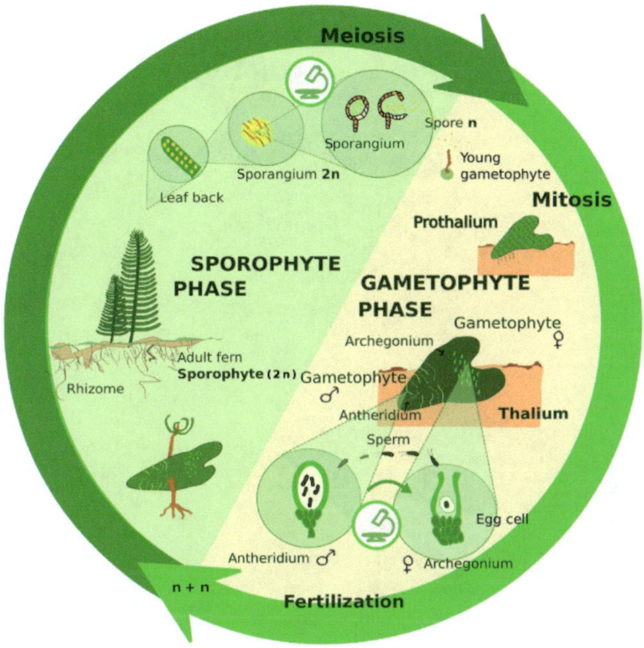

Figure 8.1: Sporophyte and gametophyte phases in Meiosis.

Source: Image by Wikipedia

Moreover, the evolution of propagules for the purpose of vegetative propagation has been established in many plant families. Zygotic embryos are considered to be resemble (superficially) such structures. Hence, both ontogeny and morphology would not be able to provide decent grounds for considering what specifically should be a plant embryo.

8.3. VARIOUS WAYS TO BECOME AN EMBRYO

In plant embryogenesis, the major and defining step is its starting from a non-embryonic precursor cell. The natural occurrence of the process begins from the fertilization of an egg cell with the transformation of a haploid gamete into a diploid sporophyte.

The event would, to the greatest possibility generate reprogramming related to the cellular state as well as set the following embryo development steps in motion.

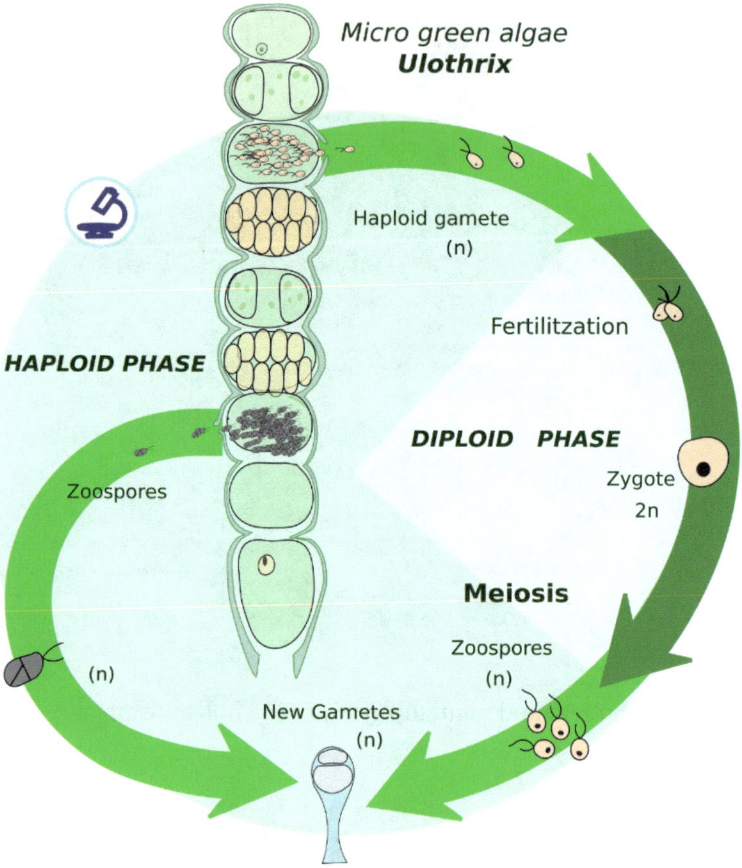

Figure 8.2: Zygotic Embryogenesis.

Source: Image by Wikipedia

However, significantly in a plant, the initiation of an embryo is restricted to merely an egg cell. Demonstration of the carrot somatic cells growing in a completely synthetic medium with the synthetic auxin phytohormone 2,4-D producing many embryos is considered as the most creative observations of the late 1950s.

The somatic cells being conditionally reprogrammed to embryogenesis later are commonly found in several species of plants from various sources and haploid microspores stresse. Thus, embryogenesis is considered as a characteristic being relatively generated in plant cells universally. It incites the query related to the regulation of the process.

Recent work, would give an idea for the necessity of reprogramming for the purpose of alterations at the chromatin level. Consequently, the embryogenic potential is usually curbed by the epigenetic tags requiring it to be removed before initiation.

Similarly, fertilization induces drastic changes in the architecture of the structure of the chromatin. Hence, epigenetic reprogramming is vital for the purpose of switching from a non-embryonic to an embryonic state, irrespective of the origin of the embryo.

Apart from repressive chromatin control, numerous positive regulators of embryo initiation have been detected, preferably, transcription factors. Interestingly, when overexpressed, embryogenesis (in a particular context) could be induced by quite a few transcription factors. However, to connect the functions of such proteins with the initiation of embryo relating to their usual functional context, there is no satisfactory evidence available yet.

Thus, theoretically, the plausibility of removing repressive chromatin marks would carry a temporary differentiated or 'competent' state of the cell that could possibly be induced by means of overexpressing transcription factors.

A cell could enter a new developmental activity, including embryogenesis, based on the hormones, stresses and various other features pertaining to the cellular state. Gene promotion is yet to see the experimental grounds towards a causal relationship between chromatin control, efficacy and embryogenesis. The cycle of events, currently, would be vague though.

There are various effective explanations for such dissimilarities. First, fundamental differences in regards to reprogramming competency exist based on the previous state of cell development. Inconsistency pertaining to the cellular point of view related to somatic cells in a tissue (epidermal cells or cambial stem-like cells) can possibly rule the limit of removing epigenetic marks.

According to another theory, neighboring cell interactions ascertain a cell's reprogramming competency. In embryonic cell cultures, stimulatory or inhibitory interactions both are supposed to have been noticed in relation to embryonic cell cultures along with identifying signal molecules intervening interactions. They include cytokinin, peptides, polyamines, oligosaccharides, complex proteoglycans and enzymes.

Evidently, even in additional basal plants, to some extent, a special periodicity would seem to exist in responding cells. It would recommend cell interactions, like lateral inhibition, to design reprogramming. Presuming, most plant tissues would contain an incessant symplast across plasmodesmatal connections, there would be genuine chance of signals passing amid cells throughout reprogramming.

Similarly, the embryo belonging to a variety of flowering plant species can suppress embryogenesis in the adjacent suspensor cells and ultimately personifying a lateral inhibition case. Precisely, single-cell level approaches would be necessary for describing and predicting reprogramming for embryogenesis and comprehending the definition of embryo at the molecular level.

8.4. EARLY EMBRYOGENESIS SHOOT-ROOT BODY PLAN

The development of three-dimensional shape of the embryo accompanied by special organs and tissues and the gene network arranging the development of the embryo are considered significantly unresolved issues. As per an explanatory aspect, there are three common phases in plant embryogenesis, wherein specific developmental and physiological events occur (Goldberg et al., 1989, Lindsey et al., 1993, West et al., 1993):

- Post fertilization pro-embryo
- Globular heart transition
- Organ expansion and maturation

Although there are large differences in the ways of embryo formation in various plant taxa (29), the universal trends would be considered alike, incredibly (Johri et al., 1992). By summarizing Capsella and Arabidopsis embryo development (Johri et al., 1992, Schulz et al., 1968, Hanstein, 1870, Soueges, 1919, Mansfield et al., 1992), the following conclusions could be drawn:

- It is regarded among the most relevantly studied forms of plant embryogenesis, going back to the conventional studies of

Hanstein, Schaffner and Soueges with Capsella (Schulz et al., 1968, Hanstein, 1870, Soueges, 1919)

- It involves a consistent pattern of division throughout the early stages allowing to trace cell heredities historically (Mansfield et al., 1992)
- According to other studies, new perceptions have been provided regarding the processes controlling the development of embryo (Meinke, 1991, Amasino, 1993).

Embryo formation would be resulting from the asymmetric cleavage of the zygote, accompanied by a suspensor as well as an embryo proper constituting specific developmental fates. The Arabidopsis and Capsella zygotes have asymmetric cellular component distribution, like the nucleus and the majority of the cytoplasm in the upper-portion of the cell. On the contrary, the middle to lower portion is controlled by a huge vacuole.

The egg cell gives rise to spatial asymmetry (Schulz et al., 1968). The zygote is asymmetrically divided into two daughter cells of specific size. It contains a small, upper terminal cell and a huge, lower basal cell, along with a polarized longitudinal axis formation inside the embryo (Raven et al., 1992, Schulz et al., 1968, Hanstein, 1870, Soueges, 1919, Mansfield et al., 1992). The past 125 years of histological studies point out that the various regions of the mature embryo would have come into existence from the terminal and basal cells (Johri et al., 1992, Schulz et al., 1968, Hanstein, 1870, Soueges, 1919, Mansfield et al., 1992). The embryo proper would be generated through the small terminal cell, forming majority of the mature embryo.

The terminal cell and embryo proper give rise to cell lineages specifying the cotyledons, shoot meristem, hypocotyl area of the embryonic related to the embryonic axis (Johri et al., 1992, Schulz et al., 1968, Hanstein, 1870, Soueges, 1919, Mansfield et al., 1992) and the radicle, or embryonic root partially (Dolan et al., 1993). On the other hand, the huge basal cell resulting from the lower part of the zygote is divided forming an extremely specialized, terminally differentiated organ of the embryo, known as the suspensor.

Figure 8.3: Cotyledon.

Source: Image by Pxfuel

The suspensor is supposed to have only 7 to 10 cells in Arabidopsis. The embryo proper is anchored by the suspensor to the embryo sac surrounding it and the ovule tissue serving as a channel for the passage of nutrients from the maternal sporophyte to the developing pro-embryo (Yeung et al., 1993). After the heart stage, the suspensor undergoes senescence and would no longer be a functional part of the seed after maturation.

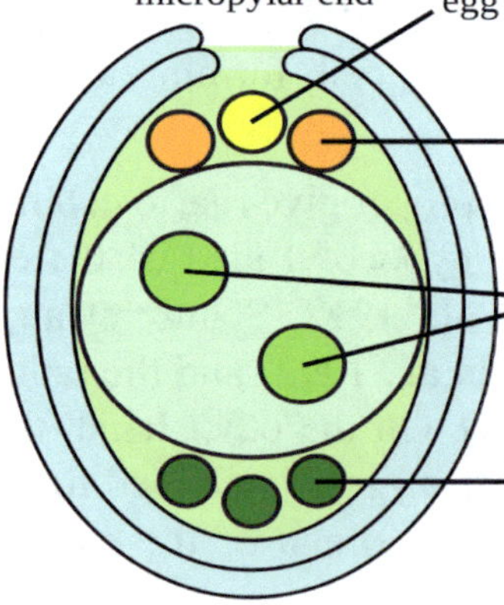

Figure 8.4: Embryo Sac.

Source: Image by Wikimedia Commons

The root meristem sees a lot of involvement from the hypophysis deriving from the uppermost cell of the suspensor. Thus, the suspensor as well as a portion of the radicle region related to the embryonic axis, would be existing from the cell lineages being derived from the basal cell.

Following the division of the zygote, it is essential for different set of genes becoming active in the terminal and basal cells. Controlling the various gene expression events early in embryogenesis, would be through the polarized organization of egg cell, the zygote or both, is yet to be studied.

For example, the pre- localized controlling factors inside the egg cell initiate a flow of events resulting in the lineage-dependent differentiation associated with terminal and basal cell derivatives. On the other hand, the *de novo* synthesis of regulatory factors spread asymmetrically to the terminal and basal cells on the first cleavage is directed by the zygotic genome following fertilization. Whatsoever, either event would result in the separate specification of the terminal and basal cells effecting from intrinsic factors instead of extracellular signals (Davidson, 1993).

8.4.1. Formation of Embryonic organs and tissue-types

After the formation of the embryo proper, two vital events are supposed to occur:

- The longitudinal- apical -basal axis regions differentiate from each other generating embryogenic organ systems
- The three primordial tissue layers related to the embryo are determined

During the globular heart transition phase, both such events occur. At the time of pro-embryo and globular stages, the embryo proper shows a spherical shape.

The first noticeable cell differentiation occurrs at the 16-cell stage while the protoderm or external cell layer of the embryo proper is generated forming the hypophysis at the suspensor top. Corresponding cell differentiation events inside the embryo proper result in producing an internal procambium layer and a mid-layer of ground meristem cells.

A radial axis of differentiated tissues inside the globular embryo is established by the spatial arrangement of protoderm, ground meristem and procambium layers. However, the hypophysis presence at the basal end of the embryo proper would develop an apical- basal polarity inside the region signifying that the globular embryo in the formal sense is not radially symmetrical.

A drastic variation related to the morphology of the embryo proper occurs immediately after the globular stage. Two lateral domains appearing on the apical end (top) establish the cotyledons. Elongation of the hypocotyl region of the axis starts and at the basal end (bottom), developing the root meristem differentiation from the hypophysis region (Dolan, 1993).

Hereafter, the embryo proper is heart-shaped showing bilateral symmetry along with the body and tissue layers of the mature embryo (and post embryonic plant) being formed. Differential cell division accompanied by expansion rates and asymmetric cleavages in several cell planes mediate morphogenic changes throughout the said period (Raven et al., 1992). In contrast to the events of migration occurring in various types of animal embryos, there is no occurrence of cell migration in plants (Davidson, 1993).

8.4.2. Embryogenesis Terminates with a Dormancy Period

A significant change occurs in the embryonic development throughout the organ growth and maturation stage. A transfer from a pattern formation activity to an accumulation activity of storage product occurs during the period, preparing the young sporophyte for embryonic dormancy as well as post embryonic development.

Cell division and expansion events result in the cotyledons and axis drastically increasing in size (Mansfield et al., 1992). Ground meristem cells inside each of the organs are extremely distinctive accumulating huge quantities of storage proteins and oils to be used as a source of food by the seedling post- germination (Mansfield et al., 1992).

During the said period, one differentiation event happening would be however, the formation of cell layers from the shoot meristem, localizing in the region of upper axis between the two cotyledons (Barton). Hence, the differentiation of shoot and root meristems opposite to the embryonic axis poles does not happen together (Dolan et al., 1993, Barton). The embryo would be reaching its maximum size, the embryonic cells and surrounding layers of the seed becoming dehydrated, ceasing of metabolic activities and beginning of the embryonic dormancy period inside the seed would occur at the end of organ expansion and maturation period (Goldberg et al., 1989, Lindsey et al., 1993, West eta l., 1993, Mansfield).

8.5. PATTERN FORMATION IN ZYGOTIC EMBRYOGENESIS

For many years, plant development biologists have been busy with the pattern formation related to angiosperm embryos. Additionally, for serving as a phylogenetic classification reference, pattern formation appeared to be well-conserved adequately. Arabidopsis studies have revealed that after the first asymmetric division, there are three more determinant steps:

- The octant stage comprising of two stages of four cells
- The protoderm formation
- The primordia initiation

The small apical cell generated following an asymmetric division does not expand throughout the following three divisions or during the successive periclinal divisions providing the cells forming the protoderm. The protoderm formation restricting cell expansion would be crucial for the rest of the developmental phases. It could be implied as result of the Arabidopsis mutant emblOl-1. Unregulated expansion of the cell in the embryo culminates the production of the enlarged cells filling up the entire seed (Meinke in De Jong et al., 1993).

Mutation of an embryo is considered to have been examined as a component of a saturating genetic screen. From the supposed 500-1000 vital genes (a number depending on the statistical extrapolations) essential for survival and positive reproduction in Arabidopsis (Jurgens et al., 1991), around 40 regulate the formation of the embryo axis pattern elements (Mayer et al., 1991).

The independent establishing of both apical-basal and radial patterns can be depicted through the mutants analysis (Jurgens, 1995). Hence, mutants contain radially organized vascular, ground and epidermal tissues, unstable in the apical-basal patterns, like gurke and fackel (Mayer et al., 1991). Similarly, lots of mutant organs at the molecular level during their elongation of cell and division plane alignment occur at the precise corresponding positions. It would thus, support the concept that expanding polarized cell and alignment of division plane is considered unnecessary for spatial development (Traas et al., 1995).

Recently it is considered to have been noticed that raspberry mutants were unable to experience the globular-heart transition the cotyledons - axis differentitions. Hence it would ascertain independent tissue differentiation occurrence, concerning patterns in embryos of higher plants. Incidentally, the globular shape of the embryo persist and losing the embryonic regions would not be affecting the tissue layer formation along the globular embryos radial axis (Yadegari et al., 1994).

In other ways, knolle mutants which would not have an epidermal cell layer, would be projecting unusual morphologies recommending the radial axis specification requirement before forming a normal apical-basal pattern (Mayer et al., 1991).

KNOLLE gene cloning and sequencing have shown the KNOLLE protein similarity with the syntaxins, belonging to the family of the protein engaged in vesicular trafficking or membrane fusion to suggest the possibility of the KNOLLE gene functioning in cytokinesis (Lukowitz et al., 1996).

8.6. MATURATION AND GERMINATION

8.6.1. Developmental Control

The accumulation of specific sets of mRNAs and equivalent proteins in the embryo and endosperm characterize the developmental stages of seed (Hughes and Galau, 1991). The storage protein accumulation is supposed to be an indicator of the maturation phase (Galau et al., 1991) following the accumulation of late embryogenesis abundant (LEA) proteins with few being revealed as ABA- inducible and assuming to participate in desiccation tolerance (Dure et al., 1989).

The seed storage protein genes expression is majorly observed in the process of developmental and genetic controls. The probability seems to be that multigene families emerged by the duplication of genes specify majority of the storage protein. The sequences of such multigene family members are considered to be developing in an independent manner (Dure and Chlan, 1981).

The developing embryos of angiosperms with 20000 RNAs, with some expressed sequentially and getting accumulating at different rates: would be both regulated at the transcriptional or post-transcriptional levels (Goldberg et al., 1989).

For example, the species that would accumulate both types of globulin, the precedence would invariably be taken over by the legumin synthesis than that of that of vicilin. Moreover, complete legume deposition would be much more compared to that of vicilin (Raynal et al., 1992). Such essential temporal difference is considered to have been seen in several dicots, like pea, grapevine, radish and soybean.

By the study of the localization of the Kunitz trypsin inhibitor (KTI), Perez-Grau and Goldberg (1989) have revealed the future of cells is specified immediately on reaching the globular stage.

The KTI3 mRNAs are supposed to accumulate at the micropylar end of the embryos towards the late globular and heart stages though not being found in other parts of the embryo. Additionally, it would be localized particularly inside the ground meristem layer of cell.

The results are an indication of the apical-basal and radial patterns concerning the globular embryo would be categorized into various regions of transcription. It demonstrates that apart from controlling temporally, proteins are also controlled spatially during the formation of seeds (Goldberg et al., 1994).

Maturation would be inclusive of physiological processes confirming the dormancy of an embryo. It involves ABA accumulation, environmental stress, inhibition of growth and quiescence maintenance. It has been noticed that precocious germination before desiccation and embryo maturation was prevented by the exogenous application of ABA (Thomas, 1993).

Therefore, ABA rising significantly throughout the late embryogenesis is regarded as an important regulator of the process and would be balancing gene expression at the minimum during the transcriptional processes of seed development (for a review, see Giraudat et al., 1994).

It is considered to have been observed through studies related to patterns of gene expression, with or without exogenous ABA that more than 150 genes in various species are ABA- inducible (Giraudat et al., 1994).

8.6.2. Genetic Control

So far, the analysis of maturation is supposed to have been done most widely at a physiological level. However, the effect of late embryogenesis -centric genes have been revealed recently, through genetic approaches.

Subsequently, in *Arabidopsis thaliana* and *Zea mays*, the genetic control related to late embryogenesis was studied by the mutant classification wherein, either ABA is not synthesized or would stop responding to the hormone.

The presence of mutants lacking ABA (biosynthesis), in Arabidopsis (the ABA mutant) and maize (the Vp2, Vp5, Vp7, Vp9 mutants) have elucidated the pathway of ABA and function in controlling the expression of seed genes.

The Viviparous mutants in maize exhibiting advanced germination with respect to the mother plant and the ABA mutants in Arabidopsis, considered to be remaining non-dormant show lower levels of ABA (Neill et al., 1986; Rock and Zeevaart, 1991).

The storage protein accumulating in Arabidopsis and maize is reduced at certain times excluding the other times. Hence, ABA levels variations are not considered as the sole developmental signal for the regulation of storage protein gene expression (Pang et al., 1988; Paiva and Kriz, 1994).

Even though in the ABA biosynthetic mutant seeds, several LEA mRNA accumulating are reduced, there is essentially no correlation between the ABA- induced gene expression and the levels of ABA (Pang et al., 1988; Finkelstein, 1993).

The developmental differences in ABA content, which might be regulating seed dormancy are revealed by such observations. However, supposedly they would not be the prime factor regulating the storage and LEA protein gene expression, although the regulatory pathways depending on ABA seems to be involved (Hughes and Galau, 1991).

The response of the mutants towards ABA is impaired and different from the ABA- biosynthetic mutants in which the endogenous ABA levels would not be reduced. Moreover, an exogenous ABA supply cannot reverse their phenotypes. By the abi3 Arabidopsis mutation (Koornneef et al., 1984) and the vpl maize mutation (Robertson, 1955), precocious germination prevents dormancy.

By means of exogenous ABA in culture, a reduced sensitivity to growth inhibition was exhibited by mutant embryos. Reduced *in vivo* accumulation of several endogenous mRNA features at the developmental stages in the seed development is due to the abi3 and vpl mutations.

The mRNAs, being discussed included globulin storage proteins, like cruciferin and napin in Arabidopsis, along with LEA mRNAs (Nambara et al., 1992; Paiva and Kriz, 1994). A little reduced abundance in the ABA and the VP mutants are displayed demonstrating the developmental factors regulating such genes apart from ABA levels alone. At the same time it shows the importance of the ABB and VP1 proteins in the regulation of such genes expression (Pla et al, 1991; Finkelstein, 1993).

According to other data, ABI3 and VP1 interact with the controlling pathways separate from ABA signaling. In reality, vpl mutants are considered to be imperfect in the accumulation of anthocyanin (this would not be the case for the ABA-deficient vp mutants) due to their inability in expressing the C1 regulatory gene, considered to be activated by VP1 or exogenous ABA (Hattori et al., 1992).

Moreover, lipid accumulation is inhibited in abi3 mutants (excluding ABA mutants), unable to break down their chlorophyll for acquiring desiccation tolerance (Finkelstein and Somerville, 1990). The molecular cloning related to the VP1 and ABB genes (McCarty et al., 1991; Giraudat et al., 1992) supports the concept of such genes that code for the transcriptional activator proteins.

The main structures of the VP1 and ABI3 proteins show an equivalent domain arrangement with specific biochemical features. Considerable sequence similarities to other proteins identified and as distinctive motifs related to DNA binding are absent. Nevertheless, various regions displaying characteristics have been identified prior in transcriptional activation domains.

The level of gene expression throughout seed development would be regulated by the ABI3 and VP1 proteins as transcriptional activators, through interaction of several transcription factors in relation to specific pathways.

The ABB locus encoding a protein that shares a similar sequence with the 2C group of serine/threonine protein phosphate, was shown by Parcy et al. (1994).

The impact of late embryogenesis inherited genes involving maturation was shown through a genetic approach. For example, in Arabidopsis, an identified mutant, known as leafy cotyledon (lee), leads to defects in the process of differentiation in cotyledons and in events specific to maturation, like accumulation of storage product, desiccation tolerance and dormancy maintenance (Meinke et al., 1994; West et al., 1994).

The transformation of the embryo cotyledons and seedlings into leafy structures, resulting from the mutations are characterized by trichromes, stomata, differentiation of mesophyll cells and deficiency in protein and lipid storage bodies.

The axis region islacking in storage organelles signifying LEC1 wild type gene functioning in both the regions concerning the embryo. Additionally, the lee embryos germinate ahead of time.

The requirement of genes for post-embryonic development would also be considered active in the later part of embryogenesis (Goldberg et al., 1994). In Arabidopsis, the fusca mutants accumulate anthocyanins within their cotyledons during late embryogenesis, not developing into mature flowering plants post -germination.

Normal embryogenesis is displayed by fusca mutants along with fus3 mutants being an exception and displaying a leafy phenotype structure. However, their lethal nature is an indication of the FUSCA genes being vital for important developmental processes as well as the accumulation of anthocyanin being just a secondary effect (Castle and Meinke, 1994).

In fact, many FUSCA genes have been CONSTITUTIVE PHOTOMORPHOGENIC (COP) fDEETIOLA TED (DET) genes alleles functioning in the light-controlled development during the germination of seed (Chory et al., 1989; Deng et al., 1992; Wei and Deng, 1992).

The COP/ DET loci products seem to be suppressing light-regulated activities of genes in the dark stimulating them in the light through light-intervened signal transduction pathway. The wild-type COP/DET alleles are active at the time of maturation as the cop/det mutant genes would have been noticed in the form of fusca embryo mutants. Thus, the plant would be prepared for life post- germination by the regulatory genes expressed at the end of embryogenesis (Golberg et al., 1994).

8.7. COMPARISON BETWEEN ZYGOTIC EMBRYOGENESIS AND SOMATIC EMBRYOGENESIS

8.7.1 Zygotic Embryogenesis

The evolution of embryogenesis is considered as a positive strategy to reproduction in higher plants. The angiosperm plant (flowering plants) life cycle is categorized into two phases; the diploid sporophytic phase and the haploid gametophytic phase (Fan et al., 2008).

The gametophyte has temporary and minor complex functions compared to the sporophyte that dedicatedly produces haploid male and female gametes (Fan et al., 2008; Reiser & Fischer, 1993; Yadegari & Drews, 2004).

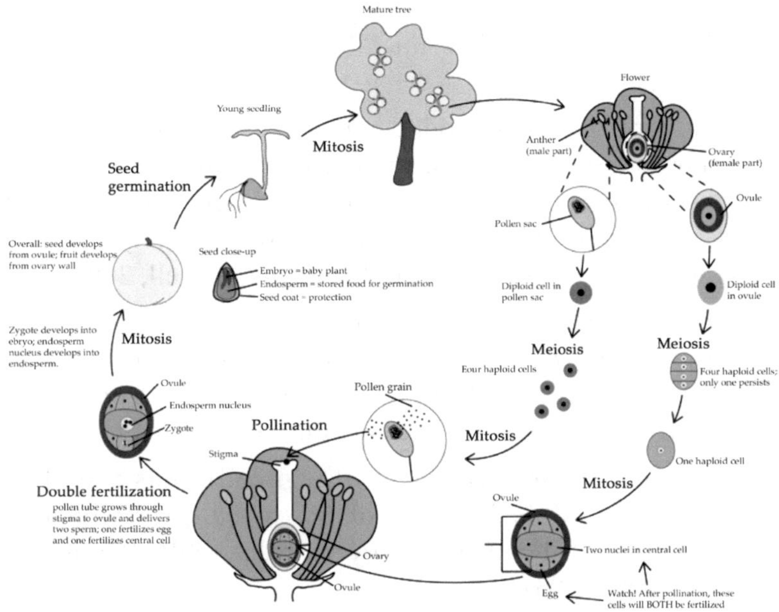

Figure 8.5: Angiosperm Life Cycle.

Source: Image by Wikimedia Commons

The development of male gametes or microgametophytes (pollen grains) occurs within the anthers forming out of a pollen mother cell via meiosis generating a tetrad of haploid cells known as microspores. At the time of maturation, on the way to form pollens, the microspore deteriorates to an asymmetric mitotic division producing two new cells: the vegetative and generative cells.

A second mitotic division is experienced by the generative cell to produce two sperms. Whereas, the vegetative cell is supposed to remain intact with the potential to produce the pollen tube that carries the sperm while grows in the female tissue of the carpel (Mc Cormick, 2004).

The embryo sac or the female gametophyte develops within the carpel comprising of three elements: the ovary, the style and the pollen grain receptacle known as stigma. One or many ovules are inside the ovary with the female gametophyte or embryo sac.

Three layers of cells encompassing the embryo sac, the nucellus and the internal and external integuments would form the ovule. The integuments are not supposed to fuse at the ovule tip providing an opening known as micropyle. The micropyle is like the "door" allowing the pollen tube to penetrate into the embryo sac.

The angiosperm sexual reproduction involves the transferring of the pollen grains from the anther to the stigma. The germination and formation of the pollen tube takes place in the stigma (Yadegari & Drews, 2004) and then moves a long way while being directed by sporophytic signals first followed by the female gametophyte (Wetering & Russell, 2004).

Later on, the pollen tube reaches the micropyle where it is directed by signals produced by the synergid cells by means of high calcium concentration (Tian & Russell, 1997). Next, is the releasing of the spermatic cells within a synergid via the filiform apparatus (Yadegari & Drews, 2004).

The fusion of one spermatic nucleus with egg cell results in the formation of the zygote (diploid) while the fusion of the second sperm with the central cell polar nuclei forms the endosperm (generally triploid), called double fertilization (Russell, 1993).

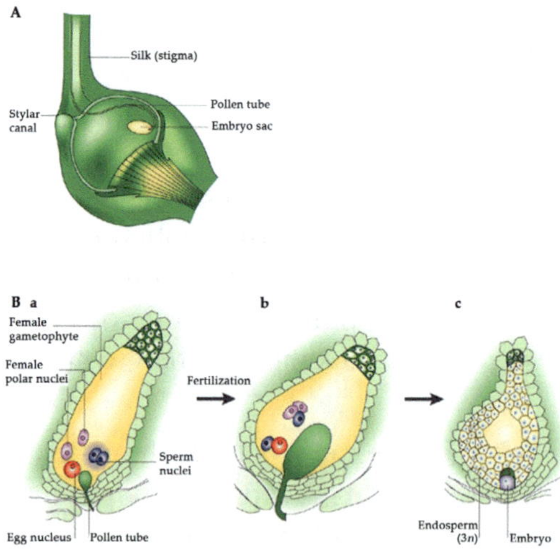

Figure 8.6: Double Fertilization.

Source: Image by Wikimedia Commons

The polarity of zygotic embryos detected, would be starting with the embryo sac being formed. The embryo sac polarity is due to the presence of a gradient of the natural auxin indole-acetic acid, moving constantly with the micropyle-chalaza axis with starting point of the expression being the micropylar region outside of the embryo sac.

By a similar pattern the haploid egg cells by the embryo post -fertilization, polarization occurs to a great extent by the nucleus being situated at the chalazal pole (Pagnussat et al., 2009). According to the reports of Pagnussat et al. (2009), the possibility of the nuclei position not being regulated by auxin at the time of forming the embryo sac, is high. Nevertheless, it funtions in the cell fate regulation during cellularization.

The subsequent diploid zygote remains extremely polarized post-fertilization. On the contrary, fusion of the other male gamete with the embryo sac central cell takes place, later developing into the triploid endosperm serving a nutritive and protective component for the embryo. In most of the plant species, the above pattern is followed by the somatic embryogenesis.

8.7.2. Somatic Embryogenesis

Somatic embryogenesis is implied as the maximum level of cell tot potency expression in plant cells. In a nutshell, the ability of a plant cell in experiencing a sequence of intricate metabolic and morphological synchronized steps for producing a whole and standard plant or sporophyte in the absence of any sexual processes, is known as totipotency.

Hence, somatic embryogenesis is related to the process of development wherein, theoretically any somatic cell would be developing into a structure resembling a zygote which would eventually become a plant (Rao, 1996; Jiménez, 2005).

Somatic embryos originate from a single cell, similar to their zygotic counterparts (Rao, 1996). Several authors have been smartly reporting the somatic embryo single cell origin. Specifically, the unicellular origin associated with somatic embryos is considered to have been reported in *Agave tequilana* (Gutiérrez-Mora et al., 2004; Portillo et al., 2007).

Generally, there are two major steps in the process of somatic embryogenesis; induction and the expression of the resultant embryos (Rodríguez-Garay et al., 2000; Gutiérrez-Mora et al., 2004; Jiménez, 2005). Theoretically, the somatic cells would be initiating the process from whichever part of the plant.

Cells considered more efficient for somatic embryogenesis commonly are the ones growing from young tissues, including immature zygotic embryos. Nevertheless, stems, roots and leaves might even be considered useful.

Generally, by easily manipulating the in *vitro* conditions, somatic embryos are induced. Growth regulator substances (GRS) are one of the vital components of the culture medium, like auxins, cytokinins, abscisic acid and gibberellins among other elements.

Moreover, not to forget, in the somatic embryogenesis process, the hormonal endogenous substances have a significant part. Among the said GRS, auxins would be the utmost vital components in the induction of the process (Dodeman et al., 1997; Jiménez, 2005; Jiménez & Thomas, 2006; Rao, 1996; Feher, 2006).

Somatic cells require the signal for cell polarization and the asymmetric division provided by auxins, similar to their zygotic counterparts (Gutiérrez-Mora et al., 2004; Pagnussat et al., 2009). Other GRS engagement is essential for balancing the hormonal elements for accomplishing somatic embryogenesis.

In the early step of development in somatic embryo, the process of induction is usually initiated by a selected auxin (the majorly used auxin for most species is 2, 4-Dichlorophenoxiacetic acid [2,4-D]) (Nomura & Komamine, 1986; Jiménez, 2005).

"The zygote", would be emulated by the early induced somatic cell. It is considered to be an extremely polarized cell with the nucleus being positioned to the furthermost part of the cell along with other extreme left vacuolated. The initial transversal cell division being asymmetrical would be generating a small apical cell as well as an extremely vacuolated basal cell.

A second round of division generates the embryo proper or two -celled pro-embryo accompanying the extremely vacuolated cell apparently being the first suspensor cell. Another third round of cell division generates an embryo head with four cells with the first cell of suspensor being affected by a second division.

Following cell divisions of suspensor cells produce the apparent hypophyseal region for the formation of the plant radicle. Successive and well-synchronized divisions result in the formation of globular stage of the somatic embryo, after the round cell divisions.

It would be vital to refer the similarities observed in most of the species for somatic embryos with a protoderm, at the globular stage. Moreover, it would be essential to mention about the suspensor not being noticed in several species, since it would not be remaining and would not be divided.

Nevertheless, such cellular and morphological idea in relation to the somatic embryo would be crucial for the radicle formation and the ultimate symmetry of the entire plant (Gutiérrez-Mora et al., 2004; Yeung et al., 1996; Supena et al., 2008).

Additionally, the morphological fate of the somatic embryos post-globular stage, would be following their genetic lineage: monocotyledonous or dicotyledonous. Forming the cotyledon primordium (cp) is a distinguishing characteristic of the dicotyledonous species.

It would be providing the traditional heart formation to somatic embryos, alike to their zygotic counterparts. On the other hand, the traditional torpedo shape is shown by the somatic embryos belonging to monocotyledonous species, which is considered to be disrupted during germination.

8.8. CONCLUSION

Embryogenesis in plants serves as an important channel between the gametophytic generation and postembryonic differentiation events occurring constantly in the shoot and root meristems in the sporophytic plant. As a result, the polarized sporophytic plan of the plant body is formed to also allow young plants to withstand unfavorable environmental conditions as well as a phase of hypogeal seed growth, would be considered to be established by plant embryos.

Plant embryos are considered to be less complex compared to their animal counterparts. However, similar developmental activities are supposed to be executed by them, like forming a three-dimensional organism with distinct regions, compartments and group of cells from a single- cell zygote. Such events would be occurring in the initial phase of the plant embryogenesis and are inadequately understood. In *Arabidopsis*, revelation of genes essential for embryogenic events, like pattern formation, cell differentiation and organ formation has started.

The exact molecular mechanisms involved in various cell lineages in the initial stage of plant embryogenesis are unknown. The events occurring inside the egg cell and the early embryo post- fertilization are undeciphered.

Hence, to comprehend molecular markers for the purpose of following the designated events happening at the time of early embryogenesis and entering into restricted networks of various regions after fertilization is vital. Though recently, there has been a huge development in comprehending the ways of plant embryo formation, yet, a lot remains to be unfolded.

REFERENCES

1. Czapik, R. and Izmaiłow, R., 2001. *Current Trends In The Embryology Of Angiosperms*. Dordrecht: Springer Netherlands, p.2.
2. de Vries, S. and Weijers, D., 2017. Plant embryogenesis. *Current Biology*, [online] 27(17), p.3. Available at: <https://www.cell.com/current-biology/comments/S0960-9822(17)30562-6> [Accessed 25 May 2020].
3. Dodeman, V., Ducreux, G. and Kreis, M., 1997. Review article. Zygotic embryogenesis versus somatic embryogenesis. *Journal of Experimental Botany*, [online] 48(313), pp.1493-1509. Available at: <https://pdfs.semanticscholar.org/099a/f2de693e0a3234cd0c74fb1a1c1afcb65c7c.pdf?_ga=2.116903622.1389556114.1589786684-472970029.1561352179>.
4. Goldberg, R., Paiva, G. and Yadegari, R., 2019. *Plant Embryogenesis: Zygote To Seed*. [ebook] p.3. Available at: <https://www.researchgate.net/publication/333704173_Plant_Embryogenesis_Zygote_to_Seed> [Accessed 25 May 2020].
5. Gutiérrez-Mora, A., González-Gutiérrez, A., Rodríguez-Garay, B., Ascencio-Cabral, A. and Li-Wei, L., 2012. *Plant Somatic Embryogenesis: Some Useful Considerations*. [ebook] p.5. Available at: <https://www.researchgate.net/publication/304496070_Plant_Somatic_Embryogenesis_Some_Useful_Considerations> [Accessed 25 May 2020].
6. Harada, J., Belmonte, M. and Kwong, R., 2010. Plant Embryogenesis (Zygotic and Somatic). *Encyclopedia of Life Sciences*, [online] p.1. Available at: <https://onlinelibrary.wiley.com/doi/epdf/10.1002/9780470015902.a0002042.pub2> [Accessed 25 May 2020].
7. Lopes, M. and Larkins, B., 1993. *Endosperm Origin, Development, And Function*. 5th ed. [ebook] p.1. Available at: <https://www.ncbi.nlm.nih.gov/pmc/articles/PMC160370/pdf/051383.pdf> [Accessed 25 May 2020].
8. Amasino et al., 1993
9. Berlath et al., 1993
10. Berlath et al., 1993
11. C. S. Gasser et al.

12. Castle et al., 1993
13. Castle, 1994
14. Chory et al., 1989
15. Davidson, 1993
16. Deng et al., 1992
17. Dodeman et al., 1997
18. Dolan et al., 1993
19. Dure and Chlan, 1981
20. Dure et al., 1989
21. Fan et al., 2008
22. Feher, 2006
23. Finkelstein, 1993
24. Forsthoefel et al., 1992
25. Gasser et al.
26. Giraudat et al., 1992
27. Goldberg et al., 1994
28. Gutiérrez-Mora et al., 2004
29. Hanstein, 1870
30. Hattori et al., 1992
31. Hughes and Galau, 1991
32. Jiménez, 2005
33. Johri et al., 1992
34. Jurgens et al., 1991
35. Koornneef et al., 1984
36. Kranz et al., 1994
37. Kriz, 1994
38. Lindsey et al., 1993
39. Lukowitz et al., 1996
40. Mansfield et al., 1992
41. Mayer et al., 1993
42. Mc Cormick, 2004
43. McCarty et al., 1991

44. Meinke et al., 1994
45. Meinke in De Jong et al., 1993
46. Nambara et al., 1992
47. Neill et al., 1986
48. Nomura & Komamine, 1986
49. Pagnussat et al., 2009
50. Paiva, 1994
51. Pang et al., 1988
52. Pla et al, 1991
53. Portillo et al., 2007
54. Rao, 1996
55. Raven et al., 1992
56. Raynal et al., 1992
57. Reiser & Fischer, 1993
58. Reiser et al., 1993
59. Robertson, 1955
60. Rock, 1991
61. Rodríguez-Garay et al., 2000
62. Russell, 1993
63. Schulz et al., 1968
64. Sheridan et al., 1988
65. Shevell et al., 1994
66. Somerville, 1990
67. Soueges, 1919
68. Supena et al., 2008
69. Thomas, 1993
70. Tian & Russell, 1997
71. Traas et al., 1995
72. Wei, 1992
73. West et al., 1994
74. Wetering & Russell, 2004
75. Yadegari & Drews, 2004
76. Yadegari et al., 1994

77. Yeung et al., 1996
78. Zeevaart, 1991

INDEX

Q

R

S

T

U

V

W

Z

Printed in the United States
By Bookmasters